# Lecture Notes in Earth Sciences 101

Editors:
S. Bhattacharji, Brooklyn
G. M. Friedman, Brooklyn and Troy
H. J. Neugebauer, Bonn
A. Seilacher, Tuebingen and Yale

**Springer**
*Berlin*
*Heidelberg*
*New York*
*Hong Kong*
*London*
*Milan*
*Paris*
*Tokyo*

Andreas Lang   Kirsten Hennrich
Richard Dikau (Eds.)

# Long Term Hillslope and Fluvial System Modelling

Concepts and Case Studies
from the Rhine River Catchment

Springer

Editors

Prof. Dr. Andreas Lang
K. U. Leuven
Redingstraat 16
3000 Leuven, Belgium
andreas.lang@geo.kuleuven.ac.be

Dr. Kirsten Hennrich
Geographisches Institut
der Universität Bonn
Meckenheimer Allee 166
53115 Bonn, Germany
kirsten.hennrich@giub.uni-bonn.de

Prof. Dr. Richard Dikau
Geographisches Institut der Universität Bonn
Meckenheimer Allee 166
53115 Bonn, Germany
rdikau@gub.uni-bonn.de

"For all Lecture Notes in Earth Sciences published till now please see final pages of the book"

ISSN 0930-0317
ISBN 3-540-00982-5 Springer-Verlag Berlin Heidelberg New York

Cataloging-in-Publication Data applied for
Bibliographic information published by Die Deutsche Bibliothek.
Die Deutsche Bibliothek lists this publication in the Deutsche Nationalbibliographie; detailed bibliographic data is available in the Internet at <http://dnb.ddb.de>.

Springer-Verlag Berlin Heidelberg New York
a member of BertelsmannSpringer Science+Business Media GmbH

http://www.springer.de

© Springer-Verlag Berlin Heidelberg 2003
Printed in Germany

The use of general descriptive names, registered names, trademarks, etc. in this publication does not imply, even in the absence of a specific statement, that such names are exempt from the relevant protective laws and regulations and therefore free for general use.

Product liability: The publishers cannot garuantee the accuracy of any information about the application of operative techniques and medications contained in this book. In every individual case the user must check such information by consulting the relevant literature.

Typesetting: Camera ready by author
Cover design: E. Kirchner, Heidelberg
Printed on acid-free paper    32/3142/du - 5 4 3 2 1 0

# Table of Contents

## I Large scale fluvial system modelling

## II Case studies from the Rhine river catchment and Central Europe

# Table of Contents

**Sidorchuck, A.,** Lab. of Soil Erosion and Fluvial Processes, Geographical Faculty, Moscow State University, 119899 Moscow, Russia

**Verstraeten, G.,** Laboratory for Experimental Geomorphology, K.U.Leuven, Redingenstraat 16, B-3000 Leuven, Belgium

**Wainwright, J.,** Environmental Monitoring and Modelling Research Group, Department of Geography, King's College London, Strand, London, WC2R 2LS, UK

**Walling, D. E.,** Department of Geography, University of Exeter, Amory Building Rennes Drive Exeter EX4 4RJ, UK

**Wasson, R.,** Centre for Resource and Environmental Studies, Institute of Advanced Studies, The Australian National University, Canberra ACT 0200, Australia

# Concepts and approaches to long term and large scale modelling of fluvial systems

Andreas Lang[1], Kirsten Hennrich[2], and Richard Dikau[3]

[1] Fysische en Regionale Geografie, K.U. Leuven, Redingenstraat 16, 3000 Leuven Belgium

[2] UFZ Centre for Environmental Research Leipzig-Halle, Riverine and Lacustrine Landscapes, Brueckstrasse 3a, 39114 Magdeburg, Germany

[3] Geographisches Institut, Universität Bonn, Meckenheimer Allee 166, 53115 Bonn, Germany

## 1 Introduction

Modelling the evolution of human–impacted fluvial systems over longer periods and larger spatial scales was the topic of a workshop held in Düsseldorf, Germany from 4 – 6 May, 2001. The workshop aimed to bring modellers with experience in the development and use of conceptual, theoretical and mathematical models for larger scales together with researchers looking at long–term system development from a more classical and empirical research perspective. The workshop provided an opportunity to exchange ideas and laid the foundations for further collaboration, and in fact for this publication.

The need for such a workshop became clear in the course of designing a large research project focusing on the evolution of the whole Rhine river system (Lang et al., 2000; 2003), within the framework of IGBP–PAGES LUCIFS (Land Use and Climate Impacts on Fluvial Systems during the Period of Agriculture). Whereas results from many detailed case studies in the Rhine catchment are already available, modelling approaches that allow integration of this wealth of data have yet to be implemented.

The need to get researchers from both ends of the spectrum – modellers and field–oriented scientists – together to fill the gap between the two approaches is, however, not unique to the work carried out in the Rhine river system. It is an issue common to much of palaeo-environmental research. Detailed information is available from drill cores, outcrops, and small drainage basin studies. But integration of these results is difficult and often possible only in a rather vague and qualitative way. In principle, it should be possible to apply mathematical modelling to quantitatively link information from different temporal and spatial scales. This would facilitate better understanding of long-term system behaviour, enable identification of key locations where additional information is needed, and provide a predictive tool for studying the effects of future global changes. In addition, contrasting model results with empirical field and historical data should allow model evaluation and refinement and thus help to improve the mathematical models. We hope that the approaches and cases studies documented here can offer new ideas and contribute towards integration of results and thus better understanding of environmental systems also in other regions and other systems. We wish that the momentum generated at the workshop in Düsseldorf will be apparent from the contributions, that it will continue as a driving force beyond the Rhine river project and that it will also help to stimulate other research initiatives elsewhere.

## 2   The research framework

The Rhine-LUCIFS project is embedded within the International Geosphere–Biosphere Programme (IGBP) PAGES Activity 5 Project LUCIFS (Land Use and Climatic Impacts on Fluvial Systems during the Period of Agriculture, Wasson et al., 1996; Wasson and Sidorchuk, 2000). It is thus part of global change research which recognises the major interactions between the Earth's physical, biogeochemical and human societal systems. A major component of the interactions between the physical and the biogeochemical system is transport of water and sediment in fluvial systems, i.e. rivers and their catchments. The transport of water, sediment, nutrients, and carbon on land and from land to the oceans is clearly of global biogeochemical significance. Inclusion of these fluxes in models of the Earth system is a goal of the IGBP (International Geosphere–Biosphere Programme), with the overt rationale is to contribute to understanding of the Earth system, and thus to achieve better land and water management in the face of land use and climate change. LUCIFS will contribute to this goal by supplying temporally and spatially resolved material budgets and flux estimates.

The above term 'Land Use Impact' can be understood to include not only agricultural activity, but also the direct impact of humans on river systems through engineering and management measures. Today, fluvial transport systems are clearly influenced by man. It is equally true that fluvial systems both now and in the past have had profound effects on human activities. Future changes in the fluvial system will affect human activity and livelihood, e.g. through flooding and associated damages, pollution of water by sediments and nutrients, and silting up of river channels and floodplains. So knowledge about the long–term sensitivity of the fluvial system to changes in the physical and biogeochemical environment is crucial to society.

The application of LUCIFS results to natural resource management, in particular to catchment management, is of high priority. This emphasis reflects the growing global problem of supplying adequate fresh water to both human populations and to aquatic ecosystems. IGBP has also recognized water as a key cross-cutting theme, along with carbon, and food and fibre.

Coming decades will certainly see continuing and indeed increasing land use change, as well as climate change, both natural and human–induced. Important questions posed by these changes are:

− How will rivers and their catchments respond to these changes?
− How will fluxes of biogeochemically active materials change?
− What will be the biogeochemical responses?
− What management is appropriate to control catchment response?

To answer such questions there is a need for accurate models that build on long–term information and are applicable to large catchments.

The purpose of this introduction is to briefly describe the research environment and provide an overview of the papers included here. This volume is divided into two parts, one conceptual/modelling part and a second part with case studies from the Rhine system and its surroundings. The first part deals with more conceptual and mathematical approaches to long–term modelling of soil erosion and catchment or landscape evolution. The second part presents case studies offering data to test and validate mathematical models. Bringing together both components is essential for better understanding of the long–term and the large-scale system behaviour and evolution.

## 3   Large scale fluvial system modelling

In the first part approaches towards conceptual and mathematical approaches for long–term modelling of soil erosion and catchment or landscape evolution are presented. The first paper 'A LUCIFS Strategy: Modelling the Sediment Budgets of Fluvial Systems' by Aleksey Sidorchuk, Desmond Walling and Robert Wasson reviews the strategy of PAGES-LUCIFS and presents modelling approaches that have been used on fluvial systems of the Russian plain. Different qualitative and quantitative strategies are presented for systems of different size. For the Khoper river basin the authors show that about 90% of the soils eroded since agriculture started have been deposited within the catchment and have not yet entered the sea.

The second paper 'Linking Short– and Long–Term Soil–Erosion Modelling' by John Wainwright, Anthony J. Parsons, Katerina Michaelides, D. Mark Powell, and Richard Brazier reviews different approaches for modelling soil-erosion starting from particle–scale process-based approaches to methodologies used for whole catchments. The authors stress that even some small flaws in the empirical equations used for modelling over shorter periods and smaller areas can lead to largely erroneous results when applying these models to longer–times and larger areas. To overcome such problems Wainwright et al. suggest an alternative framework and argue strongly for more appropriate process representation. Also, the authors state that model development must occur in parallel with parameterisation, in order to avoid the possibility of model failure due to absence of requisite data.

In 'Modelling sediment fluxes at large spatial and temporal scales' Nicholas Preston and Jochen Schmidt develop a conceptual modelling approach based on the hierarchy of landform structure. The authors start with a review of models that were explicitly designed for modelling landform evolution and models that could achieve parts of the required tasks. Then, they sketch an aggregated model structure aimed at closely matching larger-scale landscape configuration. Sediment redistribution in this approach is modelled based on the frequency/magnitude spectrum of sediment production events, temporal aggregation used in parameterisation, and sediment routing through a network of morphological landscape units. Besides being useful for modelling, this approach should also provide a framework for integrating results obtained using other methodologies.

In 'Modelling the Geomorphic Response to Land Use Changes' Anton Van Rompaey, Gerard Govers, Gert Verstraeten, Kristof Van Oost and Jean Poesen present a spatially distributed approach for calculating sediment delivery to rivers. The model was validated and calibrated based on sediment yield data from central Belgium. Land use change scenarios were extracted from historical map sequences and used as model input. The results clearly show non-linear system behavior: small changes in the size of arable land lead to disproportionate changes in soil erosion and sediment delivery.

Kristof Van Oost, Gerard Govers, Wouter Van Muysen and Jeroen Nachtergaele in 'Modelling Water and Tillage Erosion using Spatially Distributed Models' compare results from modelling runs with long-term erosion patterns inferred from soil profile truncation and medium-term erosion patterns derived from $^{137}$Cs concentrations. The study shows that soil erosion models are valuable tools for understanding long–term patterns. The cumulative effect of water erosion over periods of several thousands of years closely predicts the contemporary spatial pattern of soil truncation. A major change in erosion and sedimentation has occurred in recent decades where soil erosion is dominated by tillage. In addition the effect of changes in landscape structure on soil erosion are studied. It is shown that representation of field boundaries is extremely important when the focus shifts from field to catchment scale. Another important issue when trying to link surface processes and sedimentary records is shown in another modeling run: simulating the effects of soil erosion can help to explain changes in sediment properties.

The modelling study presented by Tom J. Coulthard and Mark G. Macklin in 'Long–term and large scale high resolution catchment modelling: Innovations and challenges arising from the NERC Land Ocean Interaction Study (LOIS)' shows how it is possible to simulate the evolution of large river catchments over Holocene time scales. The Holocene-evolution of major upland tributaries of the Yorkshire Ouse is simulated on a regular grid basis. Input parameters include the land cover history as derived from palynological studies and a rainfall record that was extracted from peat bog wetness indices. Periods of increased sediment discharge are a response to wetter periods in the rainfall record, and the magnitude of sediment yields is amplified after catchment deforestation. Differences in sediment yield between sub-catchments are caused by intermediate sediment storage and later remobilisation.

We believe that this selection of papers will provide an overview of conceptual and mathematical modelling approaches available for longer–term and larger scale drainage basin evolution.

# 4 Case studies from the Rhine river catchment and Central Europe

In the second part of this book a number of case studies are presented, exemplifying the variety of results that are already available. These studies range from very high temporal resolution of a single slope to sediment budgets for large drainage basins with much reduced temporal resolution.

The first paper of the second part, 'Large to Medium–Scale Sediment Budget Models – the Alpenrhein as a Case Study', by Matthias Hinderer presents Late Pleistocene and Holocene sediment budgets for the Rhine catchment upstream of Lake Constance. Changes in weathering regime, processes, and transport capacity are reflected in changing sediment fluxes and the build-up and reworking of specific sedimentary sinks. According to Matthias Hinderer this part of the Rhine river represents a relatively simple and largely closed sedimentary system. In the second part of the paper a concept is developed to enable construction of large to medium scale sediment budgets for sedimentologically open systems. A four-step approach is proposed that involves defining and investigating representative subsystems, upscaling this information and basing stratigraphic correlations between subsystems on deposits of large events that can be traced throughout larger parts of the river system.

From the other end of the Rhine system Hans Middelkoop and Nathalie E.M. Asselman review in 'Impact of Climate and Land Use Change on River Discharge and the Production, Transport and Deposition of Fine Sediment in the Rhine basin – a summary of recent results' the work that has been carried out in the Netherlands in the past few years. One focus of that work was on the evolution of the Rhine–Meuse delta, for which an extremely detailed description of the sedimentary architecture has been established. Due to the almost complete Holocene sedimentary record, caused by relative sea–level rise and land subsidence, many new insights into the characteristics of the Rhine river system have been gained. Amongst others these insights include the identification of channel belts and avulsion frequencies, and a recognition of the important controls on river behaviour at time scales of centuries to millennia. The other focus of the paper is the work that has been carried out on the Rhine basin as a whole and on a decadal to century time scale. This research was mainly driven by the potential impacts of future climate and land use changes on the river discharge, sediment production and transport to the delta. Several models have been developed and integrated in a GIS framework to simulate system behaviour. The results show the complexity of the system: whereas climate change may accelerate erosion rates, land use changes in the form of a decrease in arable land, lead to reduction of erosion. Larger parts of the system are also only poorly coupled: increasing erosion in the Alps has little effect on the sediment load downstream. Hans Middelkoop and Nathalie E. M. Asselman show nicely how models can be used to improve understanding of large systems and enable evaluation of scenarios for decision support.

From the other end of the spatial spectrum - on a local scale but with high temporal resolution – two case studies of historic landscape changes due to human impact are presented: 'Changing Human Impact during the Period of Agriculture in Central Europe: The Case Study Biesdorfer Kehlen, Brandenburg, Germany' by Gabriele Schmidtchen and Hans–Rudolf Bork, and 'Land Use and Soil Erosion in northern Bavaria during the last 5000 Years' by Markus Dotterweich. Both papers document excellently how detailed field work, combined with analytical results and studies of historical archives can be used as sedimentary archives for obtaining palaeo–information almost at the event scale.

Hans–Rudolf Bork and Andreas Lang present in 'Quantification of past soil erosion and land use / land cover changes in Germany' two approaches to regionalising results from case studies to larger regions. From the results it is clear that soil erosion is not only a modern problem in the Rhine river system. Phases of increased soil erosion have occurred during all phases of stronger human impact in the Neolithic, Bronze Age, and Iron Age. The maximum rates of soil erosion occurred in the medieval period. The main trigger of soil erosion seems to have been the intensity of land use, as this was the critical factor for the landscape's sensitivity to erosion. Later, during late medieval and modern times, the extent of arable land reached more or less its present distribution and results are available at a much higher resolution. For this period maxima in soil erosion can clearly be associated with high magnitude rainfall events. Extreme soil loss occurred during the first half of the $14^{th}$ century and in the second half of the $18^{th}$ century.

We are very optimistic that future modelling approaches will allow researchers to bring together knowledge, as documented in the second part of this book, to give a more coherent view of the behaviour of the Rhine river system as a whole.

## Acknowledgements

The organizers of the Düsseldorf workshop were able to support the attendance of participants through funding provided by the German National Committee on Global Change Research, the Deutsche Forschungsgemeinschaft, and the University of Bonn. We are very grateful to Wolfgang Schirmer and the University of Düsseldorf for making it possible to hold the meeting at the splendid location of Schloss Mickeln. We would like to thank the reviewers for their comments and suggestions, which have contributed considerably to the quality of the contributions. These include: Gerardo Benito, Tom J. Coulthard, Gerard Govers, Matthias Hinderer, Mark G. Macklin, Hans Middelkoop, Nick Preston, John Dearing, Jochen Schmidt, Kristof Van Oost, Anton Van Rompaey, David Favis-Mortlock, and John Wainwright. Finally, we would like to thank the authors for producing the papers for this book and for having been so patient and gentle with us throughout the publication process.

## References

Lang, A., Preston, N., Dikau, R., Bork, H.–R., and Mäckel, R. (2000): Land Use and Climate Impacts on Fluvial Systems During the Period of Agriculture  Examples from the Rhine catchment. PAGES Newsletter, 8/3: 11–13.

Lang, A., Bork, H.R., Mäckel, R., Preston, N., J. Wunderlich, and Dikau, R. (2003): Changes in sediment flux and storage within a fluvial system - some examples from the Rhine catchment.-In: Hydrological Processes (in press).

Wasson, R.J. (ed.) (1996): Land Use and Climate Impacts on Fluvial Systems during the Period of Agriculture. PAGES Workshop Report, Series 96–2, 51 pp.

Wasson, R.J and A. Sidorchuk (2000): Land Use and Climate Impacts on Fluvial Systems during the Period of Agriculture (LUCIFS). PAGES Newsletter, 8/3: 11.

*Andreas Lang*
*Kirsten Hennrich*
*Richard Dikau*

# Part I

# Large scale fluvial system modelling

# A LUCIFS Strategy: Modelling the Sediment Budgets of Fluvial Systems

Aleksey Sidorchuk[1], Desmond Walling[2], and Robert Wasson[3]

[1] Lab.of Soil Erosion and Fluvial Processes, Geographical Faculty Moscow State Univ.119899 Moscow, Russia
[2] Department of Geography, University of Exeter, Amory Building Rennes Drive Exeter EX4 4RJ, UK
[3] Centre for Resource and Environmental Studies, Institute of Advanced Studies, The Australian National University, Canberra ACT 0200, Australia

## 1  Introduction

Cumulative Global Change is occurring through the removal of forests, conversion of marginal land to cultivation, and intensification of cultivation. Systemic Global Change, in the form of changes in climate and atmospheric chemistry, is likely to alter land use patterns during the next century. All of these changes will affect rivers and their catchments, altering the fluxes of water, sediment, nutrients, carbon and pollutants. The effects of past changes of land use and climate are still being felt in many catchments, and are difficult to understand without an historical perspective. Future changes, when superimposed on changes triggered in the past, will produce complex responses, which may be difficult to anticipate.

There is therefore a clear need for a better understanding, and a better theory, of the response of fluvial systems to land use and climate change, in order to anticipate and perhaps predict future changes, and to understand current dynamics. Although most of the relevant scientific research has been conducted in small catchments, most interest in future responses relates to large catchments. The time period over which a large catchment responds to land use or climate change is much longer than that for a small catchment; much longer than most instrumental time series.

The PAGES–LUCIFS (Past Global Changes – Land Use and Climate Impacts on Fluvial Systems component of the IGBP, the International Geosphere and Biosphere Program) Project aims to assemble a library of fluvial system responses from around the world, using case studies selected to represent both a wide range of land use and catchment types, and areas where land use has evolved slowly over millennia and where industrial agriculture has been transplanted to non–agricultural landscapes during the last few centuries. Models capable of application to this library of responses are being developed to provide the tools for anticipation and/or prediction of future change.

Although the impacts of Global Change on fluvial systems are important in themselves, change to these systems will also affect the coastal zone by increasing material fluxes to the coasts in some cases and decreasing fluxes in others.

Changes in the carbon and nutrient fluxes of rivers will also impact on global biogeochemical cycles, particularly in terms of carbon and nutrient sequestration within the fluvial system and the coastal zone. Rivers play an important role as a link between the major global biogeochemical systems, but, more than that, they are crucially important to human well being and for aquatic organisms. The importance of fluvial systems, as phenomena that are affected by Global Change and in turn affect Global Change, is clear, along with their wide–ranging societal and ecological significance.

## 2 LUCIFS Aim and Scientific Questions

The main aim of the LUCIFS programme is to understand the changes in water and particulate fluxes through fluvial systems and their associated budgets, over the period of agriculture. These changes and budgets are to be investigated at global, regional and local scales, and over both long and short time scales. There are five questions, which are central to LUCIFS investigations. These are:

1. How have fluvial systems responded to past changes in climate and/or land use?
2. What are the key factors that have controlled water and particulate fluxes (as sediment, P, C and other) in different regions?
3. In each region, does the response and sensitivity of the system to these key factors vary spatially and temporally?
4. In each region, how do long–term processes affect the present day responses of fluvial systems?
5. What feedback exists between variations in water/particulate fluxes and global environmental change?

By answering these questions within a global conceptual framework, a global view can be obtained. The LUCIFS project involves temporal scanning of a complex spatial phenomenon – the fluvial system, which is controlled both by natural and socio–economic processes. However, information concerning the fluvial system and about key controlling factors can only be partial at any temporal scale. The level of information varies both spatially and across the different components of the fluvial system and between different regions. One of the main approaches to filling information gaps involves modelling of spatial phenomena and temporal processes.

## 3 A definition of the fluvial system

The fluvial system can be defined (in the broadest sense) as the complex of connected channels (including ephemeral flows on hillslopes) and reservoirs on the surface of the globe, associated with water flow and the erosion, transport and deposition of sediment and other particulate matter. According to the different types of flow (permanent or ephemeral), the geomorphological history and the types of erosion and sedimentation processes operating, the system can be characterised by different elements. For example, on the Russian Plain the most common morphological elements of the fluvial system (Figure 1) are: 1) the slopes and rills; 2) elongated gentle troughs on the slopes (lozhbina); 3) active smaller gullies and their fans; 4) aggraded larger gullies or stream channels (balka); 5) active stream channels; 6) rivers and floodplains; 7) river deltas.

**Fig. 1.** The fluvial system structure on the Russian Plain.

All these elements are interconnected and can be transformed from one to the other in space and time due to erosion or deposition. Rapid erosion and sedimentation under conditions of variable climate and sparse vegetation (often human–induced) cause rill and ephemeral gully formation on slopes, which may be transformed into active bank and slope gullies and, in favourable conditions, into large gullies and stream channels. In this situation, small rivers transport high sediment loads and can be aggraded. They often become buffering areas for sediment deposition between the slopes and large river channels. With a mild climate and good vegetation cover, erosion on slopes is very slow, and gentle elongated troughs are formed not only by surface erosion, but also by seepage water and by slow mass movement. In such situations, streams and small rivers are fed by clean water and can erode their channels.

The main methods of modelling the spatial properties of the fluvial system involve topographic and geomorphological mapping, recently aided by the construction of Digital Terrain Models. These DTMs can be used to show the whole complex of connected elements comprising the fluvial system, the elements of the system, or the morphological parameters of these elements and their statistical characteristics. Such maps allow the linear and non–linear interpolation and extrapolation of the spatial features on the basis of spatial relationships. For example, interpolation of the map of gully density between key sites on the Russian Plain was based on relationships with the density of the network of streams and small rivers and the percentage of arable land (Kosov et al., 1989).

## 4  Response of the fluvial system to long–term changes in climate and land use: a qualitative approach

The fluvial system is controlled by the well–known factors influencing erosion–deposition processes, namely: 1) precipitation (rainfall and snowmelt); 2) relief; 3) vegetation cover; 4) soil texture and erodibility. This control is complicated, and often one variable is related to several factors. At the qualitative level an increase in precipitation or relief generally increases the activity of erosion processes, whereas their decrease reduces erosion and promotes deposition. An increase in vegetation cover and soil strength decreases erosion intensity, and vice versa. As the elements of the fluvial system are connected, the same combination of controlling factors may cause sequentially altered processes of erosion and deposition through the system. These factors can be influenced by climate change or by human impact. The sign and intensity of process reactions to the same changes in controlling factors can be different in different parts of the fluvial system. For example, in the upper Don River basin the fluvial system was highly eroded in the cold Late Glacial time, when precipitation was twice as high as present , surface flow was reinforced by the low permeability of permafrost, and vegetation cover was sparse. This was a period characterised by intensive sheet and rill erosion on bare slopes, formation of long and deep gullies, and incision in the small and medium–sized river valleys. The eroded matter was mainly transported along the large rivers and accumulated in the river deltas (Figure 2a). At the beginning of the Holocene, the incidence of permafrost was reduced over this region and soil permeability increased dramatically. This resulted in a sharp decrease in surface flow, aggradation in the valleys of the small and medium–sized rivers, and intensive deposition in the large gullies. During the Holocene thermal optimum, dense forest–steppe vegetation covered the territory and the surface flow became relatively low due to infiltration and increased evotranspiration. These changes caused the general stabilisation of the whole fluvial system (Figure 2b). During the period of intensive agriculture (the last 400 years) the forests were cut and 60–70 % of the territory was ploughed. Intensive slope and gully erosion occurred and sediment concentrations in the surface runoff increased rapidly within the upstream parts of the fluvial system. Deposition occurred in the large gullies and small river channels and on the floodplains, and suspended sediment concentrations decreased downstream (Figure 2c).

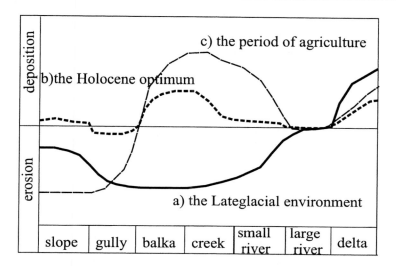

**Fig. 2.** Qualitative estimate of erosion and deposition rate change along the fluvial system of the upper Don River during a) the Late Glacial; b) the Holocene optimum; c) the period of intensive agriculture.

## 5 Response of the fluvial system to long–term changes in climate and land use: a quantitative approach

The sediment budget for the water flow can be described by the equation of mass conservation, which can be written in the simplified form:

$$\frac{\partial Q_s}{\partial X} = q_{sl} + M_0 W + M_b d - C \frac{d}{D} V_f W \qquad (1)$$

Here $Q_s = QC$ represents the volumetric sediment discharge (m³ s⁻¹); $Q =$ water discharge (m³ s⁻¹); $X =$ longitudinal co-ordinate (m); $C =$ mean volumetric sediment concentration; $q_{sl}=$ specific (for a unit of the length) sediment discharge of the lateral input from the river basin (m² s⁻¹); $M_0 =$ upward sediment flux from the river bed (m s⁻¹); $M_b =$ sediment flux from the channel banks (m s⁻¹); $W =$ flow width (m); $d =$ flow depth (m); $V_f =$ sediment particle fall velocity in turbulent flow (m s⁻¹); $D =$ the bed load particle size. The fall velocity $V_f$ in turbulent flow with a mean velocity $U$ is less than Stokes's fall velocity $V_{st}$ in laminar flow due to turbulent oscillations. Therefore, as a first approximation, the modified formula of Hwang (1983) can be used to estimate $V_f$.

$$V_f = \frac{V_{st}}{1 + \frac{0.5U}{(9.0V_{st})^2}} \qquad (2)$$

In the case of very fine particles and high turbulence, $V_f$ can be close to 0. The left side of the equation of mass conservation (Equation 1) defines the sediment budget for the channel reach. The right hand side of the equation defines the sediment fluxes: the first term is the lateral flux from hillslopes in the basin, the second is the upward flux from the river bed, the third is the sediment flux from the banks, and the fourth is the downward flux to the river bed.

A 1D equation of mass conservation is universal for a fluvial system, which is defined as a 2D net of 1D flowlines. This approach can be used at most scales, ranging from the small rills on a slope to large river basins. For some special cases (sedimentation on floodplains, in reservoirs and in lakes), a 2D mass conservation equation is coupled with 2D and 3D fluid dynamics modelling. Equation 1 is a first order ordinary differential equation, and the solution of this equation depends on the form of the terms, which describe the sediment fluxes. When $M_0 = kU(U^2/U_{cr}^2)$ and $dW_b = k_b \frac{W}{d} M_0$ (cf. Sidorchuk, 1999), Equation 1 can be written in the nondimensional form:

$$\frac{d\partial Q_s}{Q\partial X} = k_1 \frac{U^2}{U_{cr}^2} + \frac{q_{sl}d}{Q} - \frac{d}{D}C\frac{V_f}{U} \tag{3}$$

Here $U_{cr}$ is the critical velocity for erosion initiation and $k_1$ is an empirical coefficient. Equation 3 indicates that a nondimensional sediment budget is controlled by three factors : 1) the relative sediment input from the river basin; 2) the ratio between the flow velocity and its critical value (a measure of flow erosivity); and 3) the ratio of the fall velocity for the sediment particles to a flow velocity. When the right hand side of Equation 3 is positive, sediment concentration increases along the stream and the segment of the fluvial system is eroded . When the right hand side of Equation 3 is negative the system is depositional , and when it is zero the fluvial system is in equilibrium . The solution of Equation 3 shows the change in sediment concentration along the stream:

$$C_i = C_{i-1}exp\left(-\frac{V_f}{UD}(X_i - X_{i-1})\right) + \frac{UD}{V_f}\left(\frac{k_1 U^2}{dU_{cr}^2} + \frac{q_{sl}}{Q}\right)\left[1 - exp\left(-\frac{V_f}{UD}(X_i - X_{i-1})\right)\right]$$

$$\tag{4}$$

The term $\frac{V_f}{UD}$ controls deposition within the channel section. Deposition increases with the particle size of the suspended sediment, and decreases as flow velocity and turbulence increase. Part of the increase in sediment load within the reach can be accounted for by erosion of the channel bed and bank $\left(\frac{k_1 U^2}{dU_{cr}^2}\right)$ and by sediment delivered directly to the channel from the surrounding river basin $\left(\frac{q_{sl}}{Q}\right)$. Substantial deposition leads to the decreasing influence of the first term in Equation 4 on the suspended sediment load and an increase of the effect of the second term. The second term in Equation 4 mainly reflects the local sources of the sediment load. Thus the sediment budget regime of a fluvial system can be classified in the 'phase space' of three key non–dimensional numbers viz. $\left(\frac{k_1 U^2}{dU_{cr}^2}\right), \frac{V_f}{UD}$ and $\left(\frac{q_{sl}}{Q}\right)$. Several cases can be identified:

1. The flow velocity is much greater than the critical velocity for particle detachment. In this case the upward particle flux is significant. Fall velocity

$V_f$ is also high and intensive exchange of sediment between the channel and the flow will occur in the fluvial system. The erosion and deposition processes change along the channel. Deposition increases because of the lateral flux of sediment from the catchment. This is the most complicated regime of the fluvial system, and Equation 3 in its general form must be used for modelling.

2. Exchange of sediment between the channel and the flow is intensive, but upward and lateral fluxes are equal to the downward flux. This is the case of channel bed dynamic equilibrium, which is common for medium and large rivers. In small rivers, this equilibrium is often disturbed by accelerated erosion of the catchment, which results in a high lateral flux and causes aggradation in small river channels. In Equation 3 the term on the left–hand side can be omitted:

$$\frac{q_{sl}d}{Q} + k_1\frac{U^2}{U_{cr}^2} - \frac{d}{D}C\frac{V_f}{U} = 0 \qquad (3.2)$$

$$C = \frac{\frac{q_{sl}d}{Q} + k_1\frac{U^2}{U_{cr}^2}}{\frac{dV_f}{DU}} \qquad (4.2)$$

3. The flow velocity is much greater than the critical velocity for particle detachment, and the upward particle flux is significant. Flow turbulence is high and the fall velocity $V_f$ is low. Deposition is limited and erosion is intensive. Such a regime is common for active rills, for the initial stages of gully erosion and for channel erosion in cohesive sediments. In Equation 3 the last term on the right–hand side can be omitted:

$$\frac{d\partial Q_s}{Q\partial X} = k_1\frac{U^2}{U_{cr}^2} + \frac{q_{sl}d}{Q} \qquad (3.3)$$

$$C_i = C_{i-1} + \left(\frac{q_{sl}}{Q} + \frac{k_1 U^2}{dU_{cr}^2}\right)\Delta X \qquad (4.3)$$

4. Flow velocity is less than the critical velocity for particle detachment, and the upward particle flux is limited. At the same time catchment erosion delivers only fine particles in the lateral flux and the flow turbulence is sufficiently intense to reduce the fall velocity $V_f$ . This is the case of channel static equilibrium, with limited (zero) erosion and deposition. All terms in Equation 3 are equal to zero:

$$\frac{k_1 U^2}{dU_{cr}^2} = 0, \frac{V_f}{UD} = 0 \qquad (3.4)$$

$$C_i = C_{i-1} \qquad (4.4)$$

5. The same as in case 4, but with a significant lateral flux of sediment from the catchment and/or upper reaches of the channel, and aggradation of channels.

This is common for reservoirs, floodplains, and river deltas. In Equation 3 the first term on the right–hand side is omitted:

$$\frac{1}{Q}\frac{\partial Q_s}{\partial X} = \frac{q_{sl}}{Q} - \frac{1}{D}C\frac{V_f}{U} \tag{3.5}$$

$$C_i = C_{i-1}exp\left(-\frac{V_f}{UD}\left(X_i - X_{i-1}\right)\right) + \frac{q_{sl}UD}{QV_f}\left[1 - exp\left(-\frac{V_f}{UD}\left(X_i - X_{i-1}\right)\right)\right] \tag{4.5}$$

# 6    The LUCIFS modelling strategy

Three main options for modelling the sediment budget of a fluvial system can be identified viz.

1. Process–based modelling of the whole basin. Equations 3–3.5 are applied to all types of erosion and deposition processes in the basin. The spatial limits are set by DTM resolution. This approach requires large quantities of initial data, involves many model parameters, and is generally used only for small catchments.
2. Combination of simple statistical models (e.g. USLE, RUSLE and similar) for estimating slope erosion and empirical data on gully erosion, with process–based modelling in the river network. Equations of the 3–3.5 type are applied only to permanent streams. The main limitation of this approach is the estimation of the empirical delivery ratio which needs to be applied to the sediment load that is transported from the slopes and gullies to the rivers. The data requirements for the calculations can generally be met and the approach can be used for long–term modelling of fluvial systems of different sizes.
3. Combination of simple statistical models for estimating erosion rates with empirical information on erosion, transport and deposition in the river net. The main limitation of this approach is the need to estimate the empirical sediment delivery ratio for all catchments within the basin. This approach is best suited to coarse modelling.

## 6.1    Modelling of small catchments

Use of the GULTEM model (Sidorchuk and Sidorchuk, 1998) to model gully erosion and thermoerosion provides an example of the first of the approaches listed above. This model describes the first stage of rapid gully development. During snowmelt and/or rainfall events, flowing water erodes a rectangular channel in the topsoil or in the gully bottom. Shallow landslides quickly transform the gully cross – section to a trapezoidal form during the period between water flow events. GULTEM was applied to the net of flowlines established using a topographic DEM. The soil texture was estimated for each horizon (including the surface horizon with vegetation cover) with a different composition. Runoff due to snowmelt and rainfall was calculated from meteorological information using a physically–based hydrological model.

DEMs were used to provide information on elevations, flowline directions and gradients, and catchment areas. The original interactive procedure was extended to incorporate the filling or linking of closed depressions associated with errors in the interpolation of the initial relief. One of eight possible directions of flow with maximum gradient was used for flow path estimation. The ability to set the preferred direction was used to estimate the influence of small features such as roads or the pattern of ploughing. The catchment area of any point was calculated as the sum of the pixel areas of all flowlines linked to the point above it. Elevations of the soil surface were estimated from DEMs for each soil with a similar texture, and the main soil characteristics were determined from direct field observations.

The main meteorological and hydrological processes taken into account in modelling surface runoff were:

- Precipitation amount, in form of snow or rainfall.
- The thermal dynamics of the snow, the thawing of the snow and the melt-water output.
- Interception of water by crops and natural vegetation.
- Water storage in micro–depressions.
- Infiltration

The runoff was represented using a kinematic wave approach coupled with the Manning formula, applied to the network of flowlines. The width and depth of flow was calculated using empirical regime formulae. The main formula used to describe gully erosion was of the 3.3–4.3 type, where erosion is the main process and the eroded material is completely removed from the system.

GULTEM was used to model gully development on the Yamal Peninsula, in the presence of deep permafrost, snowmelt and rainfall. One such gully, for which both initial and actual longitudinal profiles were available, is situated on the right bank of the Se–Yakha River in the central part of the peninsula (Sidorchuk, 1996a). Before 1986 there was a shallow linear depression with a dense vegetation cover and ephemeral flow. In 1986 an exploitation camp was built at the top of the basin. Surface disturbance and increased meltwater flow led to intensive gully erosion. A gully 840 m long (measured along the thalweg ) developed. In 1991 and 1995 the longitudinal profile of the gully was investigated. The initial profile was available from a large–scale map. The depths of runoff associated with snowmelt and rainfall were calculated using available meteorological data. The coefficients in Equation 3.3 were calibrated using observations for the 1986–1995 period.

The results of the calculations (Figure 3) are mainly controlled by the soil texture (aggregate size and cohesion), vegetation cover density (a function of land use) and water discharge (a function of the climate).

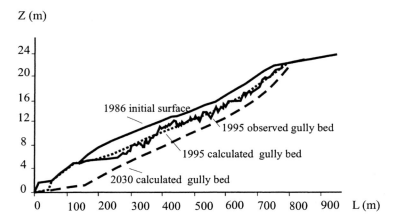

**Fig. 3.** Calculated gully dynamics on the Yamal Peninsula (Northern West Siberia).

## 6.2 Modelling medium–sized basins

The reconstruction of sediment budgets in medium–sized fluvial systems can be based on the second approach: i.e. the combination of integrated (lumped) estimates of the sediment supply to the river network from the slopes, rills, gullies and other ephemeral channels with accurate evaluation of the sediment budget in the distributed permanent river network. The basin must be divided into subcatchments; each connected to a segment of a stream or small river. The volume of erosion from the slopes, rills and gullies must be calculated for each subcatchment for a given period of time. The Sediment Delivery Ratio (SDR) must be estimated for each slope–rill–gully complex. The volume of erosion combined with the value of SDR gives the value for the lateral flux in Equation 3 for each stream or river segment. Other terms in Equations 3 and 4 are calculated based on regional data on channel and floodplain processes. The solution of Equation 4 gives the change in sediment discharge along the river network over a given period of time.

### A case study of the Zusha River basin

The basin of River Zusha (a tributary of the upper Oka River) is located in the Central Russian Upland, with altitudes in the range 140–280 m above mean sea level. The fluvial system here consists of slopes; rills; lozhbinas; gullies and fans; balkas; streams; and small and medium size rivers with floodplains. Erosion takes place on the upper parts of the slopes, and sedimentation occurs on their lower parts, with a predominance of erosion overall. The gullies are primarily characterised by erosion. The balka and stream valleys are the main areas of sediment deposition. Sedimentation also occurs on the river floodplains. A complicated process of sediment exchange between the bed and the flow takes place in the small and medium rivers. The contemporary rate of sheet and rill erosion for agricultural land was calculated by Belotserkovskiy et al. (1991) using two main Soil Loss models, which were verified for Russian Plain conditions viz. the State Hydrological Institute Model (SHI – Model) for estimating erosion during spring snowmelt and the Universal Soil Loss Equation for the period with rainfall. The estimated rate of soil loss varies from 3.0 to $10.0\,\mathrm{t\,ha^1\,year^{-1}}$ within the basin. The volume of gully erosion (i.e. the volume of gullies more than 50 m long) during the period of intensive agriculture was calculated by Kosov et al. (1989), who reported a mean value of $640\,\mathrm{t\,ha^1}$. The structure of the main channel network (tributaries more than 10 km long) and the main morphometric and hydrological parameters used in the calculations were derived from Hydrological Survey data.

The regional version of Equation 4 is (Sidorchuk, 1996b):

$$C = \left(C_0 - \frac{2.8Q_0S}{q_w\,(Y+1)} - \frac{C_w}{Y} - \frac{0.0025\sqrt{Q_0S}}{q_w\,(Y+0.5)}\right)\left(\frac{Q_0}{Q}\right)^Y + \frac{2.8QS}{q_w\,(Y+1)} + \frac{C_w}{Y} + \frac{0.0025\sqrt{QS}}{q_w\,(Y+0.5)} \quad (4r)$$

Here $Q_0$ and $C_0$ are the water discharge and volumetric sediment concentration in the channel flow at the beginning of the reach, $S$ is the water surface slope, $q_w$ is the lateral specific (for a unit of the channel length) water discharge in $\mathrm{m^2\,s^{-1}}$, $C_w$ is the volumetric sediment concentration in the lateral flow, and $Y = (q_{w} + V_f W)\,/q_{w}$.

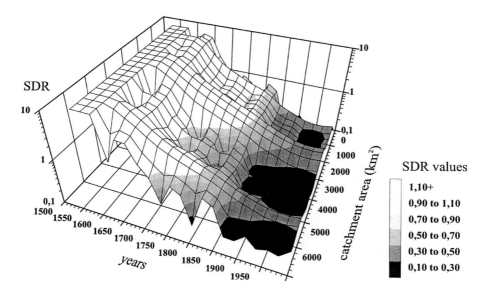

**Fig. 4.** Sediment Delivery Ratio change through time and in space for the Zusha River basin.

Sediment delivery ratios and sediment yield variations for the Zusha River catchment were calculated through the time and across space (Figure 4) for different levels of human impact. During the 16th century, under natural conditions with a very low level of slope and gully erosion, the SDR was greater than 1.0 for the entire basin. Channel erosion dominated. The sediment yield was equal to the transport capacity of the river channel, and it exceeded the input of sediment from the slopes and from gullies. As human impact at the beginning of 17th century increased and 14 % of the river basin was tilled, the SDR became less than 1.0 for the main part of the basin, because the transport capacity of the river flow (highest for the last 500 years due to maximum precipitation) remained smaller than the input of the sediment from the slopes and from the gullies. However, this input was not very great, and the rate of sedimentation in the upper reaches of the channel was also not very high. About 20–50 % of the eroded sediment was transported out of the system. During the 1930's when 71 % of the, river basin was tilled, the value of SDR was less than 0.2 for the whole basin. Only 10 % of eroded sediments was delivered to the outlet of the system. Erosion from the slopes dominated sediment yields in the upper part of the channel network and in the gullies, and was mainly controlled by land use: i.e. change in the area of arable land through time. The sediment yield in the lower river reaches was dominated by the transport capacity of the river flow, which was primarily a function of the climate (water discharge).

## 6.3   Modelling of large basins

The sediment budgets of large fluvial systems can be reconstructed using the third method, namely the Sediment Delivery Ratio approach. The large basin must be subdivided into a system of subbasins. The following values must be calculated for each subbasin for a given period of time:

— The volume of erosion from the slopes, rills and gullies.
— The Sediment Delivery Ratio defined as the ratio of the sediment transport from the subbasin outlet to the total erosion on the basin slopes by rills and gullies.

This gives:

— The sediment yield from each sub–basin outlet.
— The volume of sediment deposition within the sub–basin i.e. the difference between total erosion and the sediment yield at the sub–basin outlet.
— The depth of deposition in the river network.

When applied to the whole basin, this procedure gives the distribution of eroded sediment within the basin, the deposition depth in the network of streams and rivers, and the sediment flux at the outlet of the large basin.

## Case study: the Khoper River basin

Erosion on the slopes was calculated by Belotserkovskiy et al. (1983) using the same two soil loss models as employed in the Zusha River study outlined above; i.e. the SHI model for spring snowmelt and the USLE for rainfall. The principle of superposition of erosion factors was used for mapping of the erosion rate. Each factor in the USLE or SHI–model was mapped separately. The erosivity factor $R$ was calculated for all meteorological stations in the Khoper River basin using rainfall intensity measurements. A regional relationship between erosivity $R$ and mean annual rainfall depth was established and erosivity was estimated for the stations where only rainfall depth measurements were available. On the basis of these data, the $R$ factor was mapped using isolines. The soil factor $K$ was calculated directly from the USLE nomograph for all categories of soil texture, structure, permeability and organic matter content for soils, shown on a 1:1,000,000 soil map of the USSR. The vegetation cover factor $C$ was calculated from long–term crop rotation statistics, available for the smallest administrative districts. The seasonal changes in vegetation cover were taken into account. The same spatial resolution was used for the management factor $P$, which was estimated from regional statistics on land management. The most complicated approach was that used for the calculation of the relief factor $LS$. The national DEM was not available at the time of calculations, so the Khoper River basin was subdivided into several morphologically similar units, based on a geomorphological typology, using large–scale topographic maps. Measurements of relief characteristics were performed separately for cultivated and uncultivated land at 300–600 points, to obtain the distribution of the $LS$ factor within the unit and its mean value. Superposition of soil type and morphological units gave the main features of the soil erosion map of the Khoper River basin (Figure 5). For each of these categories, the mean soil–loss and its distribution curve were estimated.

The erosion map was used to calculate the soil–loss from the slopes for all of the catchments where sediment yield was measured by the State Hydrological Survey for a period of more than 10 years. The SDR was calculated for each of these catchments within the Khoper River basin as the ratio of the volume of the measured sediment yield from the catchment to the calculated volume of erosion $E$ in the catchment: SDR= $T/E$. These data provided a basis for establishing a regional relationship between the Sediment Delivery Ratio and river basin area (Figure 6). This relationship was extended to the whole river network of the Khoper River basin. The volume of erosion, the SDR and, accordingly, the volume of sediment transported from the basin outlet and the volume of sediment deposited within each catchment, can be estimated for any subbasin for a given time period. Information on the length of all the permanent streams in this basin is available from the measurements of the State Hydrological Survey, and the floodplain width can be estimated from the regional relationship: $W = 14.8\sqrt{Q}$ . This provides an estimate of the area of the floodplains associated with the permanent streams, and therefore the main areas of sediment deposition, and permits estimation of the depth of sedimentation on the floodplains for a given period of time (Table 1).

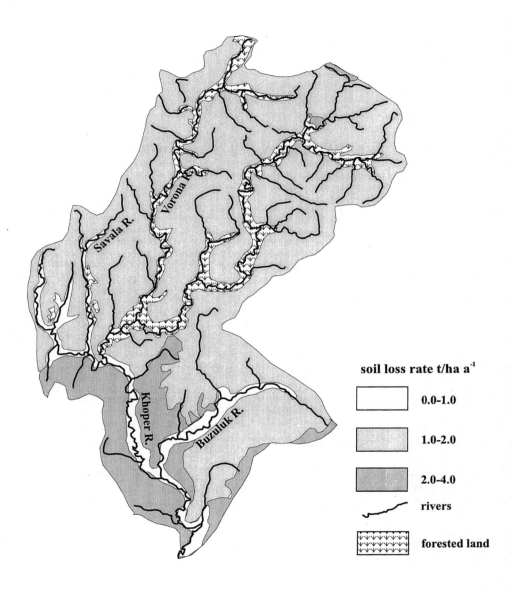

**Fig. 5.** Soil loss in the Khoper River basin, calculated with the USLE and SHI (after Belotserkovskiy et al., 1983).

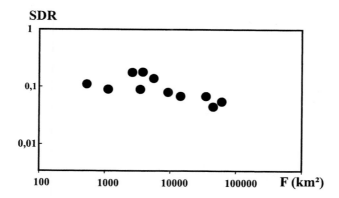

**Fig. 6.** Sediment Delivery Ratio versus basin area for the Khoper River basin.

**Table 1.** River net structure and distribution of the annual sediment deposition in the Khoper River basin (up to Novokhopersk hydrological station)

| River length (km) | < 10 | 10–25 | 26–50 | 51–100 | 101–200 | 201–500 | 501–1000 |
|---|---|---|---|---|---|---|---|
| Rivers number | 735 | 131 | 35 | 16 | 3 | 1 | 1 |
| Total length (km) | 2283 | 1961 | 1264 | 1141 | 414 | 454 | 656 |
| Total area of channel and floodplain (km$^2$) | 7.9 | 58.8 | 164.3 | 353.7 | 248.4 | 449.4 | 1377.6 |
| Total deposition ($10^6$ t/a) | 2.25 | 0.35 | 0.17 | 0.14 | 0.1 | 0.14 | 0.03 |
| Deposition thickness (mm/a) | 285 | 6.0 | 1.0 | 0.4 | 0.4 | 0.3 | 0.02 |

# 7   Conclusion

LUCIFS aims to develop an improved understanding of the variations in water and particulate fluxes through fluvial systems at various times since agriculture began on our planet. We wish to know how fluvial systems have responded to past changes in climate and/or land–use, what factors controlled fluxes of water and particulates (sediment, particulate nutrients and carbon), how sensitivity to these factors varies in space and time, and how present day changes are affected by long–term processes and trends. Finally, LUCIFS wishes to contribute to our understanding of the feedbacks to global environmental change from changes in fluvial systems. Such feedbacks occur principally through changes in the carbon cycle, modulated by sediments and nutrients delivered to the coastal zone by rivers.

Material budgets are central to all LUCIFS studies. The great variability of fluvial system types and of the level of investigation of their history, means that a single modelling framework cannot be identified. Different combinations of process–based sediment budget models, simpler statistical models and empirical relationships may be applied to LUCIFS case studies, according to the availability of data. The gaps in the databases are generally increase with the basin size and modelling period. A general simplification of the approach with an increase in basin size and the duration of the modelling period introduces objective limitations on Global reconstruction within the LUCIFS strategy. The main limitation to global reconstructions using the LUCIFS approach are the need for application of models for use in large basins, and the reality of modelling long time periods.

# References

Belotserkovskiy, M. YU., Docudovskaya, O. G., Kiryukhina, Z. P., Larionov, G. A. and Mirgorodskaya, N. N. (1983): Kolichestvennaya otsenka erozionnoopasnykh zemel' basseyna Dona (The quantitative estimate of the erosion–prone land at the Don River Basin). In: Chalov. R.S. (ed). Erosiya pochv i pusloviye prosessy, Izd. Moskovskogo Universiteta, pp. 23–41 (in Russian).

Belotserkovskiy, M. YU., Dobrovol'skaya, N.G., Kiryukhina, Z. P., Larionov, G. A., Litvin, L. F., and Patsukevich, Z. V. (1991): Erozionnyye protsessy na Evropeyskoy chasti SSSR, ikh kolichestvennaya otsenka i rayonirovaniye (Erosion processes in the European USSR, a quantitative and spatial assessment). Vestnik Moskovskogo Universiteta Series 5, Geography 2: 37–46 (in Russian).

Hwang, P.A. (1983): Fall velocity of particles in oscillating flow. J. Hydraul. Eng., 111(3): 342–351.

Kosov, B. F., Zorina, E. F., Lyubimov, B. P., Moryakova, L. A., Nikol'skaya, I. I., and Prokhorova, S. D. (1989): Ovrazhnaya eroziya (Gully Erosion). Izd. Moskovskogo Universiteta, 168 pp (in Russian).

Sidorchuk, A. (1996a): Gully erosion and thermo–erosion on the Yamal peninsula. In: Slaymaker, O. (ed): Geomorphic Hazards. John Wiley, New York, pp. 153–168.

Sidorchuk, A. (1996b): Sediment Budget Change in the Fluvial System at the Central Part of the Russian Plain Due to Human Impact. In: IAHS Publication No. 236, Erosion and sediment Yield: Global and Regional Perspectives: 445–452.

Sidorchuk, A., and Sidorchuk, A. (1998): Model for estimating gully morphology. In: IAHS Publication No. 249: 333–343, Wallingford.

Sidorchuk, A. (1999): Dynamic and static models of gully erosion. Catena, 37: 401–414.

# Linking Short– and Long–Term Soil–Erosion Modelling

John Wainwright[1], Anthony J. Parsons[2], Katerina Michaelides[3], D. Mark Powell[2], and Richard Brazier[1]

[1] Environmental Monitoring and Modelling Research Group, Department of Geography, King's College London, Strand, London, WC2R 2LS, UK
[2] Department of Geography, Leicester University, University Road, Leicester, LE1 7RH, UK
[3] School of Geographical Sciences, Bristol University, University Road, Bristol, BS8 1SS, UK

## 1   Introduction

Soil erosion by overland flow is a significant process over large areas of the Earth. It leads to specific forms of landform development over both short and long time scales. In some cases, the landscape can be dramatically modified in a matter of hours, as a result of an extreme storm event. Understanding soil erosion is therefore fundamental in being able to explain the geomorphology of these areas. The soil is also a fundamental resource for human food supplies, and its loss means direct and indirect impacts on sustainability. Off–site effects of erosion can be significant both for pollution, particularly when chemical fertilisers and pesticides have been used, and for siltation of reservoirs and other structures. In extreme cases, persistent erosion can lead to a total loss of productivity, leading to desertification. The understanding of soil erosion therefore also has important practical implications.

In order to address these implications, as well as geomorphic explanations, the use of soil–erosion models has become increasingly prevalent. Models may be used as investigative tools, explanatory tools, or predictive tools. Often, they are applied as preliminary investigations so that field resources can be more effectively directed. It is our proposition in this chapter that all of these approaches are necessary, but also that we must understand the interactions between them. A useful model should not be restricted to just one of these contexts. In order for this to be the case, we suggest that a sound process base is necessary for model development, and should be used to provide fundamental underlying tests of the adequacy of models.

## 2   Modelling at Very Small Spatial (and Temporal) Scales – The Process Base

At very small scales, it is possible to use approaches based on the movements of individual soil particles. Fundamentally, all erosion processes depend on two sets of control. First, there must be a process of entrainment, or detachment, of sediment particles. Secondly, the transporting medium – in this case water – must move the sediment particle a distance away from its original position. When considering soil erosion by the action of water, the process of entrainment is carried out in two different ways. At the start of the rainfall event, before overland flow is generated, the action of raindrop impact on the ground surface is important in the detachment of particles. The controls on the amount of detachment are essentially the kinetic energy of the rainfall, the slope, irregularity and moisture content of the surface, and particle characteristics such as size and shape (e.g. Quansah, 1981; Gilley and Finkner, 1985; Govers, 1991; Torri and Poesen, 1992). In effect, the balance is between the amount of available energy relative to the energy required to move the particle and the effectiveness of the transfer of that energy to the particle. As overland flow begins, the detachment effect of raindrops is reduced, as the raindrop energy is dissipated through the film of water at the surface (Torri et al., 1987). However, spatially this effect is highly variable as the depth of flow typically varies over a wide range due to the spatial heterogeneity of factors such as infiltration rate, microtopography, vegetation and stone cover (Abrahams et al., 1988a). Progressively, as the flow energy builds up, it may become sufficiently competent to detach particles itself. It is usually assumed that once flow detachment occurs, the ground surface is reorganized by the formation of rills that concentrate flow paths. Thus, on any one surface during a storm, there is likely to be a continuum of raindrop and flow detachment going on at any point of time.

Transport of sediment also occurs in two ways according to the presence of flow. At the start of the storm, some of the detached sediment will be transported by rebounding rain droplets in a process known as splash or rainsplash. (Note the distinction between splash and raindrop detachment.) This process occurs at relatively slow rates because of the lack of a concentrated transport medium, and is most effective at moving particles in the sand–size range. Apart from large drops resulting from processes such as leafdrip, most natural rainfall can only move particles of less than about 12 mm in diameter (Wainwright, 1992; Abrahams et al., 1988b; Kotarba, 1980; Kirkby and Kirkby, 1974; Moeyersons and de Ploey, 1976; Parsons et al., 1991). The distance moved is typically much less than a metre for fine sand, and tends to be negative-exponentially distributed away from the point of impact (Mosley, 1973; Savat and Poesen, 1981; Riezebos and Epema, 1985; Torri et al., 1987). Coarser particles will tend to creep at rates on the millimetre scale. Once overland flows are generated, they will start to transport detached particles. In unconcentrated flows, the travel distances for individual particles are still typically low, and of the order of several tens of centimetres, even for fine sands (Parsons et al., 1993). The distribution of travel distances is usually negative exponential or gamma (Wainwright et al., 1995). Concentrated flows are generally sufficiently competent to transport pebbles on hillslopes, and travel distances can be of the order of tens of metres (Poesen, 1987).

As noted above, it is usually only possible to describe the effects of the entrainment and transport processes as distribution functions. The reason for this limitation is due to the variability in sediment, morphological and flow characteristics in any real–world setting. Soils are a mixture of particles of different size, shape, density and chemistry. The effects of varying the relative positions of different particle sizes while holding the other characteristics constant can lead to dramatic changes in entrainment conditions, as demonstrated in the example in Figure 1. The morphology of most hillslopes is highly irregular, at least on the centimetre scale (often referred to as the microtopography). Turbulence in flows causes there to be instantaneous increases and decreases in the velocity of flow in a relatively unpredictable manner. These *individual* sources of variability are compounded by interactions among them. For example, flow velocities will change as surface roughness changes. Surface roughness is controlled by particle–size effects and changes in the surface morphology (form roughness). Particle sizes will obviously change as a result of selective transport, and the form roughness will be affected by concentrated erosion. Thus, it should not be anticipated that there are simple ways of characterizing the erosion process and its development at this scale.

Nevertheless, this approach can be used successfully when appropriate simplifications are used. Wainwright et al. (1999) used the approach to simulate the redevelopment of disturbed desert–pavement surfaces. They assumed that the soil could be described as a mixture of six size classes, each of which moved according to distribution functions based on the median characteristics of the size class. Their results demonstrated that the rate of evolution, particle–size characteristics of the surface and gross erosion rates were well predicted using an uncalibrated set of parameters. Furthermore, the approach is not necessarily limited to short time scales. By coupling a particle–based simulation with a time–averaged approximation to matrix erosion, Wainwright (1991; 1994; Wainwright and Thornes, 1991) were able to demonstrate the impact of erosion processes on archaeological sites over periods of several millennia. These impacts were consistent with observations from excavated sites (e.g. Gascó et al., 1995).

# 3   Modelling at the Hillslope Scale

The above–mentioned problems have tended to limit the general applicability of particle–based techniques. Most approaches to the modelling of hillslope–scale erosion therefore simplify the situation significantly. The most widely used approach is the Universal Soil–Loss Equation (USLE) developed by the United States Department of Agriculture. USLE is an empirical model based on a multiple regression of tens of thousands of years of plot data, largely derived from sites in the US Mid–west. This underlying database is one reason for the popularity of the model – it is commonly assumed that because the model is based on such a large amount of data, that it must be correct. However, the transferability of the approach has been questioned not only to other parts of the world, but even to other regions of the US. For example, the RUSLE (Revised USLE) form

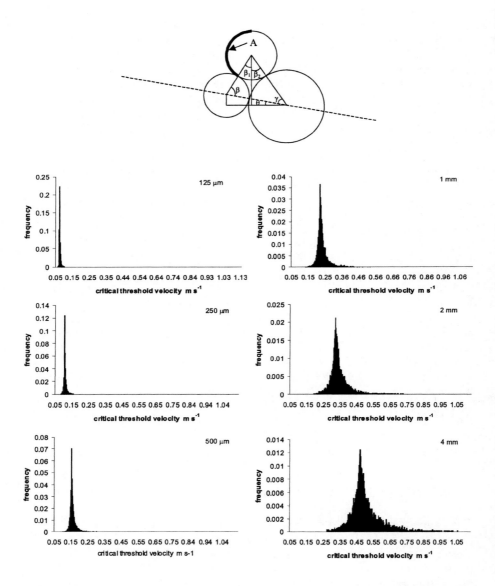

**Fig. 1.** Theoretical impact on critical threshold velocity of changing the relative sizes of a particle to be detached and its supporting grains in overland flows: a. geometrical representation (heavy shading on particle to be moved represents projected surface area); and b. distribution of threshold velocities for a constant particle size, based on random changes in the sizes of the two supporting particles.

of the model has been developed to solve problems with the application of the approach to rangelands and other environments within the US, and there are a number of variants that form ad hoc solutions to specific problems. Another reason for the enduring popularity of the model is its simplicity. Erosion rates ($A$ in $\mathrm{Mg\,ha^{-1}\,a^{-1}}$) can be predicted as a function of six variables:

$$A = R\,K\,LS\,C\,P \qquad (1)$$

$R$, the rainfall erosivity; $K$, the soil erodibility; $L$, the slope–length factor; $S$, the slope–angle factor ($L$ and $S$ are usually combined); $C$, the cropping factor; and $P$, the conservation practice factor.

Because simple nomograms have been produced that relate the different variables to other factors such as soil texture, organic matter content and infiltration characteristics, USLE is simple to apply with a minimum amount of data. However, the transferability of the functions underlying these nomograms is often severely limited. Another limitation is that the USLE is essentially designed for specific field–sized plots, that are often inappropriate for settings away from the original design area. The full extent of this particular limitation will be discussed in more detail below. Predictions are also only made at the annual scale. Furthermore, it is incomplete in terms of its process representation, in that it only represents processes of unconcentrated overland–flow erosion. It has been demonstrated in a number of settings (e.g. Morgan, 1995; Govers and Poesen, 1988) that rill and gully erosion are far more important volumetrically, so this is a significant deficiency. RUSLE has been developed to overcome some of these problems principally by including modifications to the slope–length factor $L$, to account for conditions in which rilling occurs (Renard et al., 1991). However, the fact remains that this approach is still based on empirical conditions, and does not account for the variety of mechanisms that lead to the acceleration of soil erosion.

In order to address these problems, the USDA decided to produce a process–based soil–erosion model, which it named WEPP (Water Erosion Prediction Project). WEPP accounts for the different processes involved in erosion – i.e. that due to splash, unconcentrated and concentrated overland flows. However, it does make some simplifications with regard to the latter process, most specifically relating to the process dynamics, that significantly limit the broader applicability of the model. The model has coupled climate and runoff–generation models so that event–based erosion can be predicted, and predicts the spatial pattern of erosion along a one–dimensional hillslope profile. Current tests of the WEPP model suggest that it is only about as good as the USLE at erosion prediction on the annual scale, and has high levels of uncertainty associated with the annual predictions.

Despite the use of an underlying process base, this model is essentially still an optimized fit to field data, albeit a far more elaborate one (Nearing, in press). The field data in this case have been generated from rainfall–simulation experiments, and so aim to cover a wide range of conditions in a wide range of environments, although still restricted to the US. Thus, the limitations of trans-

ferability still need to be addressed. In contrast with the WEPP model, the
European soil–erosion model, EUROSEM (Morgan et al., 1998) was designed
more specifically with a process–based approach. However, the implementation
of the process structure within the model from the outset is still no guarantee of
performance (Parsons and Wainwright, 2000). In essence, it is still necessary to
follow a hybrid approach where field and laboratory experiments are necessary
to provide refinements to the model structure and component interactions at the
same time as the collection of parameterization data.

Coupled hydrology–erosion models have the potential advantage of process–
based understandings of the erosion predictions ("getting the right answer for
the right reason" – Grayson et al., 1992). However, their structure has the disad-
vantage of propagating errors, for example in parameter measurement, through
the prediction process. Because of the highly non–linear character of the erosion process and the underlying flow hydraulics, this problem means that any
error in the prediction of hydrological parameters is significantly amplified in
the prediction of the resulting erosion (Wainwright and Parsons, 1998). The implications of this result are that significant amounts of effort should be spent
in improving the techniques of parameter estimation at a point and spatially,
and that error characteristics of predictions should be assessed and quoted in all
instances. Improved erosion models will only be developed when more detailed
spatial information becomes available for their testing, combined with the use
of model structures that are conservative with respect to the propagation of any
errors produced.

# 4  Links to the Catchment Scale

To provide process–based erosion predictions at the catchment scale, it is imper-
ative that effective estimates of the hydrological linkages within the catchment
system can be effectively assessed. Most hydrological models assume that an
entire catchment, including the coupling between the hillslope and the channel
system can be adequately represented by a uniform grid structure (e.g. SHE,
TOPMODEL). This approach may be appropriate for first–order systems, but
is poor at capturing the complexity of most fluvial systems. Alternatively, the
catchment is simplified into a series of planes, representing hillslopes, and one–
dimensional channel elements (e.g. KINEROS and the catchment versions of
WEPP and EUROSEM). KINEROS2 (Woolhiser et al., 1990; El–Shinnawy,
1993; Smith et al., 1995a; 1995b) aimed to improve the representation of dynamic interactions between hillslopes and channels by including the floodplains
in the channel cross–section. However, it has significant limitations as both the
left and right floodplains are lumped into a single component and the interactions between the floodplain and the adjacent hillslopes are ignored.

The hydrological model developed by Michaelides (2000; Michaelides and
Wainwright, 2002) attempts to overcome these limitations by incorporating a
dynamic representation of the hillslope and channel interactions, specifically for
semi–arid environments. The model has an integrated raster– and vector–based

representation of the catchment system (Figure 2), in which the hillslope and floodplain components of the catchment system are allowed to interact dynamically. The hillslope and floodplain elements are initially routed as overland flow on a two–dimensional grid with spatially distributed parameters. The channel system is made up of a series of cross sections, between which water is routed using the one–dimensional kinematic wave approximation. At low flows, the interaction is only affected by the amounts of flow that reach the edge of the channel belt. As the channel flow increases, the floodplains and lower parts of the hillslopes are progressively inundated, and the flow routed in a compound–channel system.

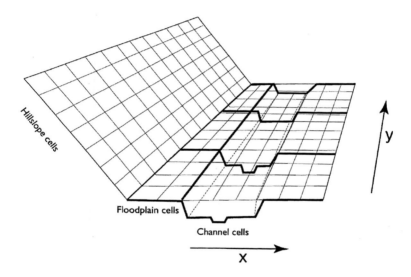

**Fig. 2.** Schematic view of catchment representation in the model of Michaelides (2002) as a series of channel cross–sections, representing the channel flow process, linked to a raster–based representation of the hillslope and floodplain conditions.

Sensitivity analyses of this model on hypothetical catchments with differing degrees of topographic coupling (i.e. the extent to which hillslopes and channels are directly linked) demonstrate complex patterns of responsiveness to parameter variability dependent on the catchment configuration, relative rainfall to infiltration rates, and parameter variability in the different hillslope and floodplain areas (Michaelides and Wainwright, 2002). Topographic coupling may not be the most important hydrological control in some catchments, depending on the specific distributions of hydrological parameters, and the nature of the barrier between the hillslope and the channel. An application to a specific catchment in south–east France also demonstrated that the model is able to represent the complex interactions that occur during a major flood event (Michaelides, 2000). Thus, our understanding of erosion patterns needs to be based on a thorough understanding of the hydrological pathways and linkages within the catchment system. Only by integrating models that deal with the complexity of the hydrological processes that drive the system can reliable erosion predictions ultimately be made.

## 5  Why the Longer Term is Important

A further significant problem with traditional hillslope–based approaches to erosion prediction becomes apparent when the results are extrapolated over much longer time scales. Parsons et al. (forthcoming) and Wainwright et al. (2001) discuss the implications of extrapolating the estimated mean erosion rates for Europe and the US, that are derived from the plot data that underlie the USLE–type approach. The mean erosion rates predict that the continents would be razed to ground level within a million years at most, a fact that is clearly incompatible with their obvious longevity. Alternative explanations, such as accelerated erosion rates due to human activity or soil exhaustion, either do not expand the life span for long enough or are incompatible with the geomorphology of the continents.

The explanation usually used to overcome this paradox is the sediment–delivery ratio (SDR; Walling, 1983). The SDR is an empirical factor relating the proportion of sediment being delivered to a specific point in the catchment system to the total amount of erosion predicted as occurring within the catchment system. There is usually a decline in the SDR with increasing catchment size, so that the largest rivers are typically delivering between 5 % and 10 % of the total predicted erosion to their outlets. However, there is a problem in using this approach, in that long–term SDRs must approach a value of 100 %, or the catchment system would progressively fill with sediment (Graf, 1988). In the few reported cases of measured SDRs of greater than 100 %, they have been reported as being erroneous (e.g. Maner, 1958) or can be related to other erosional processes, such as those relating to the presence of glaciers within a catchment (e.g. Church and Slaymaker, 1989). Furthermore, the idea of progressively increasing sediment storage is again incompatible with the observed geomorphology. Thus, as with many *ad hoc* solutions to scientific problems, the SDR actually causes more problems than it solves. Fundamentally, this approach to catchment–scale erosion estimation needs serious revision.

## 6  Reconciling the Different Approaches

An answer to this problem lies in the current incompatibility between the way erosion is measured(and thus typically modelled) and our process understanding. Plot measurements of erosion are generally normalized according to the size of the plot being studied to give a rate per unit area (e.g. $Mg\,ha^{-1}\,a^{-1}$). The implication is that, as with runoff reaching the plot boundary, the sediment delivered is eroded from all of the plot. However, the data on travel distances noted above implies that in interrill conditions, the sediment reaching the outflow is derived from the bottom few tens of centimetres of the plot. Thus, the rate per unit area is a false representation of the process. It is the flux (e.g. $Mg\,a^{-1}$) that should be measured.

Parsons et al. (forthcoming) demonstrate that in the simplest case of uniform entrainment, transport and grain size, the flux will reach a steady state once the effect of any boundary conditions and the median travel distance of particles have been reached. On plots longer than this distance, calculating by area means that effectively the constant flux is divided by an increasingly large area, so that the areal rate becomes progressively smaller. Such an observation is compatible with the observation that erosion rates measured in this way are inversely related to plot size (Evans, 1995; Rejman et al., 1999; Wilcox et al., 1996). Relaxing the assumption of constant entrainment and transport, but assuming downslope hydraulic changes characteristic of constant rainfall and infiltration, Parsons et al. were also able to show that areal rates should increase to a peak several metres below the plot boundary, before declining rapidly (Figure 3). Again, these results are compatible with field observations under equivalent conditions, where rates start to decrease about fifteen metres from the upslope plot boundary (Parsons et al., 1996). Thus, linear scaling of plot–based measurements to larger areas or longer time scales is not appropriate. Similar limitations in the assumption of linear scaling are obtained in the analysis of rill erosion, which suggests that in the absence of supply limitation, erosion rates and fluxes will progressively increase downslope. These results are also compatible with field observation (e.g. Desmet and Govers, 1997).

If the process observations and long–term rates are compatible, it should also be possible to reconcile erosion fluxes with the apparent redeposition represented by the SDR. Wainwright et al. (2001) presented an analytical solution based on the distribution functions of erosion and transport distance. In the simplest case of uniform slope conditions, it is possible to demonstrate that the effective SDR is:

$$SDR = \frac{E\left(f_e\right) V \rho_s \int_0^l 1 - F_t\left(l - x\right) dx/t}{E\left(f_e\right) V \rho_s/t} \tag{2}$$

where $E(\cdot)$ is the expectation operator; $f_e$ is the p.d.f. of the entrainment probability; $V$ is the volume of available sediment for transport $(m^3)$; $\rho_s$ is the sediment density $(kg\,m^{-3})$; $l$ is plot length (m); $x$ is the distance along the slope (m); $F_t$ is the c.d.f. of the transport distance (m); and $t$ is time (s).

In equation 2, the numerator represents the proportion of sediment travelling to the plot outlet, and the denominator represents the total sediment flux from the slope. Simplifying this relationship gives:

$$SDR = 1 - F_t(l) \tag{3}$$

It can thus be seen that the observed SDR is independent of the entrainment probability in this case. If it can be assumed that the transport distance is

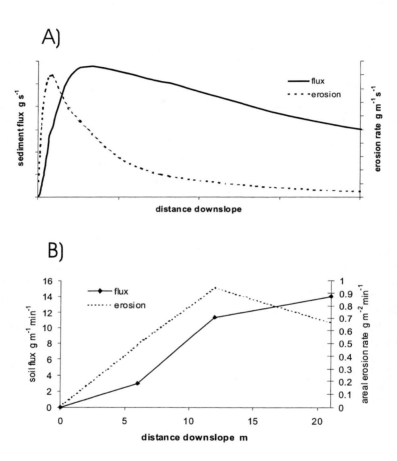

**Fig. 3.** (a) schematic diagram showing the relationship between erosion measured as a sediment flux, erosion measured as an areal rate and plot size, as calculated analytically by Parsons et al. (forthcoming); and (b) field measurements of erosion fluxes and rates on rainfall–simulation plots on grassland at Walnut Gulch, Arizona (based on data from Parsons et al., 1996).

exponentially distributed, then the following simple relationship is derived:

$$SDR = e^{-\lambda l} \tag{4}$$

where $\lambda$ is the inverse of the median travel distance $(\text{m}^{-1})$.

The form of this relationship is shown in Figure 4, and is consistent with field observations made in simple catchments (e.g. Ferro and Minacapilli, 1995). In large catchments, the apparent SDR will be a function of the variable entrainment functions and variable travel–distance functions in different parts of the catchment (e.g. hillslopes will typically have shorter travel distances for the same size particles than channels). These sources of variability probably explain the more commonly fitted power–law relationship between SDR and catchment area.

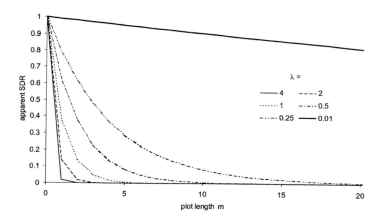

**Fig. 4.** Apparent SDR calculated by Wainwright et al. (2001) assuming constant slope conditions.

In summary, the use of a process–based understanding of erosion mechanics is fundamental in that it removes the inconsistencies of erosion predictions made for larger areas and for long time scales. Furthermore, its ability to explain the apparent pattern in SDR without predicting unfeasible amounts of sediment storage in catchment systems means that it provides a more coherent methodology for erosion prediction.

## 7 Implications and Conclusions

Erosion models can only be effectively used over large spatial scales or over longer time scales if they are able to get the right answer for the right reason.

In this paper, we have demonstrated that there are significant flaws with the empirical and semi–empirical model technology in use at the present time, and have presented an alternative framework that provides a series of fundamental predictions that are compatible with short– and long–term observations of erosion rates as well as the spatial scale of measurement.

The implications of these observations are that we must concentrate on appropriate process representations. This implication does NOT necessarily imply that physically based models are required as long as the underlying process base is sufficiently sound. Thus, predictions should be possible without the very intensive computational resources necessary for physically based predictions. Indeed, it is unrealistic to assume that a physically based erosion model can be developed to operate at the hillslope scale in the near future, despite the recent advances in computational fluid dynamics that would provide the underlying hydraulic information.

In terms of the ways in which we can expect to advance soil–erosion studies, it is important to remember that we must confront our different sources of data at all stages. If they are incompatible, it is important to find out why they may be so, rather than the usual approach of developing empirical calibrations. Only in this way can we hope to improve our understanding of the underlying process. Similarly, it is important to work between different spatial and temporal scales and between different process regimes (e.g. to avoid the artificial divide between hillslope– and channel–based approaches to catchment sediment–yield estimation). Related to this issue is the problem that process and form linkages are too frequently disregarded as an important source of information to further our understanding. If the form of the landscape is not correctly predicted by our process models, then they must clearly be wrong!

Furthermore, it must be recognized that model development and appropriate parameterization techniques must be developed together. This suggestion is a consequence of the ways in which model structure can develop to the extent that former data–collection strategies may become incomplete or even totally redundant. A model that does not have suitable data to run it, at least as a first approximation, is worthless, as it is not possible to determine why the output may deviate from that expected. Modelling is not the "cheap option" that is commonly assumed, in relation to expensive and time–consuming field measurements. Once parameters have been collected, it should not be assumed that they are inviolate or invariable. It is entirely possible that modelling can highlight the problems with a measurement technique and suggest the ways in which it might be improved (see, for example, Parsons et al., 1996). All measurements are made with error, and it is important to be able to quantify this error and its impact on the modelling process. In some cases, it may be possible to adjust the model structure in order to minimize the impacts of these errors.

A simple rule of thumb should be to be wary of all *ad hoc* solutions – even those enshrined in the literature – and empirical model calibrations. There is no reason to believe they will work elsewhere, and even if they appear to, they do not provide a solution. It is far better to work with a sufficiently complete model that accounts for differences of scale in parameters and validation data than to recalibrate inadequate models continually, providing no insight or understanding. Models of soil erosion will only develop successfully by a continued, critical appraisal of our process understanding, model–algorithm development, and data availability. We ignore the interactions of these three factors at our peril.

### Acknowledgements

Work on travel–distance approaches to sediment transport is supported by NERC grant GR3/12754 (to AJP, JW and DMP). Development of the coupled hillslope–channel hydrology model was partly supported by a Department of Geography, King's College London studentship to KM, and by NERC grant GR3/12067 (to JW). Roma Beaumont kindly drew Figure 2. We would like to thank Gerard Govers for comments on an earlier draft of this paper, although this paper does not necessarily reflect his opinions on all the issues raised.

# References

Abrahams, A.D., Parsons, A.J., and Luk, S.–H. (1988a): Hydrologic and sediment responses to simulated rainfall on desert hillslopes in southern Arizona. Catena, 15: 103–117.

Abrahams, A.D., Luk, S.–H., and Parsons, A.J. (1988b): Threshold relations for the transport of sediment by overland flow on desert hillslopes. Earth Surface Processes and Landforms, 13: 407–419.

Church, M., and Slaymaker, O. (1989): Disequilibrium of Holocene sediment yield in glaciated British–Columbia. Nature 337: 452–454.

Desmet P.J.J., and Govers, G. (1997): Two–dimensional modelling of the within-field variation in rill and gully geometry and location related to topography. Catena 29: 283–306.

El–Shinnawy, I.A. (1993): Evaluation of Transmission Losses in Ephemeral Streams. Unpublished PhD. Thesis, Department of Civil Engineering, University of Arizona.

Evans, R. (1995): Some methods of directly assessing water erosion of cultivated land – a comparison of measurements made on plots and in fields. Progress in Physical Geography 19: 115–129.

Ferro, V., and Minacapilli, M. (1995): Sediment delivery processes at basin scale. Hydrological Sciences Journal 40: 703–717.

Gascó, J., Carozza, L., Fry, S., Fry,R., and Wainwright, J. (1995): Le Laouret et la Montagne d'Alaric à la fin de l'Âge du Bronze, un hameau abandonné entre Floure et Monze (Aude). In: Guilaine, J. (ed.): Temps et Espace dans le Bassin de l'Aude du Néolithique à l'Âge du Fer, 19–22, Centre d'Anthropologie, Toulouse.

Gilley, J.E., and Finkner, S.C. (1985): Estimating soil detachment by raindrop impact. Transactions of the American Society of Agricultural Engineers, 28: 140–146.

Govers, G. (1991): Spatial and temporal variations in splash detachment: a field study. In: Okuda, S., Rapp, A., and Linguan, Z. (eds): Loess: Geomorphological Hazards and Processes, Catena Supplement 20, Cremlingen: 15–24.

Govers, G., and Poesen, J. (1988): Assessment of the interrill and rill contributions to total soil loss from an upland field plot. Geomorphology 1: 343–354.

Graf, W.L. (1988): Fluvial Processes in Dryland Rivers. Springer–Verlag, Berlin.

Grayson R.B., Moore, I.D., and McMahon, T.A. (1992): Physically based hydrologic modelling. 2. is the concept realistic? Water Resources Research 28: 2659–2666.

Kirkby, A.V.T, and Kirkby, M.J. (1974): Surface Wash at the Semi–Arid Break in Slope. Zeitschrift fr Geomorphologie, Supplementband 21: 151–76.

Kotarba, A. (1980): Splash Transport in the Steppe Zone of Mongolia. Zeitschrift für Geomorphologie, Supplementband 35: 92–102.

Maner, S.B. (1958): Factors affecting sediment delivery rates in the Red Hills physiographic area. Transactions of the American Geophysical Union 39: 669–675.

Michaelides, K. (2000): The effects of hillslope–channel coupling on catchment hydrological response in Mediterranean areas. Unpublished PhD Thesis, University of London.

Michaelides, K., and Wainwright, J. (2002): Modelling the effects of hillslope–channel coupling on catchment hydrological response. Earth Surface Processes and Landforms, 27:1441–1457.

Moeyersons, J., and de Ploey, J. (1976): Quantitative Data on Splash Erosion, Simulated on Unvegetated Slopes. Zeitschrift für Geomorphologie, Supplementband 25: 120–131.

Morgan, R.P.C. 1995 Soil Erosion and Conservation. (2nd ed.). Longman, Harlow.

Morgan, R.P.C., Quinton, J.N., Smith, R.E., Govers, G., Poesen, J.W.A., Auerswald, K., Chisci, G., Torri, D., and Styczen, M.E. (1998): The European Soil Erosion Model (Eurosem): A dynamic approach for predicting sediment transport from fields and small catchments. Earth Surface Processes and Landforms 23: 527–544.

Mosley, M.P. (1973): Rainsplash and the Convexity of Badland Divides. Zeitschrift für Geomorphologie, Supplementband 18: 10–25.

Nearing, M. (in press): Soil erosion and conservation. In: Wainwright, J., and Mulligan, M. (eds): Environmental Modelling: Finding Simplicity in Complexity. John Wiley and Sons, Chichester.

Parsons, A.J., and Wainwright, J. ( 2000): A process–based evaluation of a process–based soil–erosion model. In: Schmidt, J. (ed.): Soil Erosion – Application of Physically Based Models. pp. 181–198, Springer Verlag, Berlin.

Parsons, A.J., Abrahams, A.D., and Luk, S.–H. (1991): Size characteristics of sediment in interrill overland flow on a semiarid hillslope, Southern Arizona. Earth Surface Processes and Landforms 16: 143–52.

Parsons, A.J., Wainwright, J., and Abrahams, A.D. (1993): Tracing sediment movement on semi–arid grassland using magnetic susceptibility. Earth Surface Processes and Landforms 18: 721–732.

Parsons, A.J., Wainwright, J., and Abrahams, A.D. (1996): Runoff and erosion on semi–arid hillslopes. In: Anderson, M.G., and Brooks, S.M. (eds): Advances in Hillslope Processes. Volume 2, pp. 1061–1078, John Wiley and Sons, Chichester.

Parsons, A.J., Wainwright, J., Powell, D.M., and Brazier, R. (forthcoming): A new conceptual framework for measuring soil erosion by water. Earth Surface Processes and Landforms (submitted).

Poesen, J. (1987): Transport of rock fragments by rill flow. In: Bryan, R.B. (ed): Rill Erosion, Catena Supplementband 5, pp. 35–54, Catena, Braunschweig.

Quansah, C. (1981): The effect of soil type, slope, rain intensity and their interactions on splash detachment and transport. Journal of Soil Science 32: 215–224.

Rejman, J., Usowicz, B., and Dębicki, R. (1999): Source of errors in predicting silt soil erodibility. Polish Journal of Soil Science 32: 13–22.

Renard, K.G., Foster, G.R., Weesies, G.A., and Porter, J.P. (1991): RUSLE: Revised Universal Soil Loss Equation. Journal of Soil and Water Conservation 46: 30–33.

Riezebos, H.T., and Epema, G.F. (1985): Drop shape and erosivity. Part II: splash detachment, transport and erosivity indices. Earth Surface Processes and Landforms 10: 69–74.

Savat, J., and Poesen, J. (1981): Detachment and transportation of loose sediments by raindrop splash. Part I: the calculation of absolute data on detachability and transportability. Catena 8: 1–17.

Smith, R.E., Goodrich, D.C., and Quinton, J.N. (1995a): Dynamic, distributed simulation of watershed erosion: the KINEROS2 and EUROSEM models. Journal of Soil and Water Conservation 50: 517–520.

Smith. R.E., Goodrich, D.C., and Woolhiser, D.A. (1995b): KINEROS, A KINematic runoff and EROSion model. In: Singh, V.P. (ed.): Computer Models of Watershed Hydrology. pp. 697–732, Water Resources Publication, Fort Collins, Colorado.

Torri, D., and Poesen, J. (1992): The effect of soil surface slope on raindrop detachment. Catena 19: 561–578.

Torri, D., Sfalanga, M., and del Sette, M. (1987): Splash detachment: runoff depth and soil cohesion. Catena 14: 149–155.

Wainwright, J. (1991): Erosion of Semi–Arid Archological Sites: A Study in Natural Formation Processes. Unpublished PhD Thesis, University of Bristol.

Wainwright, J. (1992): Assessing the impact of erosion on semi–arid archaeological sites. In: Bell, M., and Boardman, J. (eds): Past and Present Soil Erosion. pp. 228–241, Oxbow Books, Oxford.

Wainwright, J. (1994): Erosion of archaeological sites: implications of a site simulation model. Geoarchaeology 9: 173–202.

Wainwright, J., and Parsons, A.J. (1998): Sensitivity of sediment–transport equations to errors in hydraulic models of overland flow. In: Boardman, J., and Favis–Mortlock, D. (eds): Modelling Soil Erosion by Water. pp. 271–284, Springer–Verlag, Berlin.

Wainwright, J., and Thornes, J.B. (1991): Computer and hardware modelling of archaeological sediment transport on hillslopes. In: Lockyear, K., and Rahtz, S. (eds): Computer Applications and Quantitative Methods in Archaeology 1990. pp. 183–194. BAR International Series 565, Oxford.

Wainwright, J., Parsons, A.J., and Abrahams, A.D. (1995): Simulation of raindrop erosion and the development of desert pavements. Earth Surface Processes and Landforms 20: 277–291.

Wainwright, J., Parsons, A.J., and Abrahams, A.D. (1999): Field and computer simulation experiments on the formation of desert pavement. Earth Surface Processes and Landforms 24: 1025–1037.

Wainwright, J., Parsons, A.J., Powell. D.M., and Brazier, R. (2001): A new conceptual framework for understanding and predicting erosion by water from hillslopes and catchments. In: Ascough II, J.C., and Flanagan, D.C. (eds): Soil Erosion Research for the 21st Century. Proceedings of the International Symposium, pp. 607–610, American Society of Agricultural Engineers, St Joseph, Mi.

Walling, D.E. (1983): The sediment delivery problem. Journal of Hydrology 65: 209–237.

Wilcox, B.P., Newman, B.D., Allen, C.D., Reid, K.D., Brandes, D., Pitlick, J., and Davenport, D.W. (1996): Runoff and erosion on the Pajarito Plateau: observations from the field. New Mexico Geological Society Guidebook, 47th Field Conference, Jemez Mountain Region, pp. 433–439.

Woolhiser, D.A., Smith, R.E., and Goodrich, D.C. (1990): KINEROS, A Kinematic Runoff and Erosion Model: Documentation and User Manual. USDA – Agricultural Research Service, ARS–77, 130 pp.

# Modelling sediment fluxes at large spatial and temporal scales

Nicholas Preston[1] and Jochen Schmidt[2]

[1] Landcare Research, Private Bag 11-052, Palmerston North, New Zealand
[2] Landcare Research, P.O. Box 69, Lincoln, New Zealand.

**Abstract.** A conceptual approach to the representation of long–term regional sediment fluxes is presented. Existing models that directly or indirectly address this objective are reviewed, and various issues associated with large scale modelling of sediment fluxes are identified. The approach offered here is designed to operate at large temporal and spatial scales. It is hierarchical, and provides a framework for integrating the results of various other approaches. The approach is based on the thesis that it is the development of landscape configuration that should be modelled on larger scales, and therefore an aggregated modelling approach for landform structure and sediment redistribution is required. Emphasis is placed on (a) frequency/magnitude spectra as a means of integrating sediment generating processes of different nature, (b) temporal aggregation for parameterisation of driving forces, and (c) routing of sediment through a topological network of morphologically defined landscape units. A series of issues requiring further research are identified, including (1) definition of geomorphic storage units, (2) establishment of effective driving and controlling factors within aggregated phases of landform development, (3) establishment of frequency/magnitude spectra for sediment generating and distributing processes, and (4) form/process/form coupling.

## 1 Introduction

Humans have influenced their physical environment through widespread deforestation, agricultural land uses and urbanisation. This has occurred over varying timescales, although for most regions human influence began at least several thousand years ago, i.e. within the late Holocene. One of the principal geomorphic effects of human-induced environmental changes has been the generation of sediment through a range of erosional processes, and the subsequent transport and deposition of those sediments. This mobilisation and redistribution has important implications for the sustainability of land use, for the design of engineering infrastructure, for contamination of waters and aquatic environments (through sedimentation itself and as a result of associated particulate fluxes), for landform development, and for prognoses of all of these phenomena under changing climate and land use scenarios. Of considerable interest, therefore, is a greater understanding of the relative significance of climatic and anthropogenic influences on the behaviour of geomorphic systems. This is particularly important as we attempt to manage and to modify our activities so as to avoid, or at least minimise, adverse impacts on the landscape. The need to determine the relative contributions of climate and land use in influencing the condition and behaviour of the physical environment is thus a contemporary research issue, and is addressed by the LUCIFS programme (Land Use and Climatic Impacts on Fluvial Systems, Wasson, 1996). The LUCIFS programme seeks to elucidate the relative roles of climate and land use on the behaviour of fluvial systems, and includes an emphasis on the flux of sediments and other particulate matter generated by agriculture. The focus is long term, i.e. covering the whole period of agriculture. Within the Rhine catchment, this is a period of ~7,000 years (Lang et al. 2000; in press). Similarly, the spatial scale is large, focussing on continental scale sediment redistribution within large fluvial systems. This requires not only a reconstruction of past landscape behaviour and an understanding of its development, but also the use of modelling approaches for the prediction of future trends. Thus three aspects for research can be identified:

1. Reconstruction of *past* sediment redistribution within fluvial systems and over the period of agriculture.
2. Identification of the relative contributions of climate and land use to the sediment flux.
3. Development of a conceptual basis for modelling the behaviour of the sediment flux, which will in turn contribute to development of prognoses of the *future* behaviour of the sediment flux.

The first aspect of this research agenda has been extensively addressed. There exists already a body of empirical data describing sediment fluxes for various localities and over different historical periods. Recent advances in the use of sophisticated dating techniques have greatly enhanced the temporal resolution of stratigraphic reconstruction (e.g. Lang and Nolte, 1999). Thus, while the key requirement identified by Dikau (1999) of empirical data collection may never be fully satisfied, it is certainly well in hand, and further research should not be greatly constrained by a lack of empirical data. Advances in dating techniques have also facilitated the development of sedimentation chronologies. This, in combination with historical geographic research, geoarchaeology and the use of climatic proxies (e.g. Glaser et al., 1999; Schmidt, 2001), has meant that considerable progress has also been made in addressing the second aspect of the research agenda (e.g. Lang, in press).

This paper focuses on the third aspect identified above - specifically, the development of an approach for modelling of sediment fluxes. Where and when is sediment mobilised? How far is it transported? Where and when is sediment deposited? What is its residence time in a given location? There are numerous models available for modelling the processes by which sediment is generated and transported, and others that to an extent predict the morphological implications of this. These are briefly reviewed, and the extent to which they are capable of addressing the questions just stated is assessed. It is questionable whether there yet exists a rigorous concept for modelling the redistribution of sediment through large systems over long periods, and an alternative approach is postulated here - one that focuses not so much on individual processes, but on aspects of geomorphic systems that control the behaviour of the sediment flux. The purpose of this paper is thus to introduce a conceptual approach to the modelling of sediment fluxes, and to identify critical issues that will need to be resolved.

## 2   The Sediment Flux System

While many human activities are local in their occurrence, their impacts - both direct and especially indirect - are often expressed at regional and continental scales. These are, accordingly, the spatial scales with which LUCIFS is concerned. Human impacts first began some thousands of years ago in most parts of the world, and the temporal scale of interest is thus the late Holocene. The processes of sediment generation and redistribution that are active within this spatio-temporal framework are manifold. These include hydrological and sedimentological processes(rainsplash, rilling, gullying, subsurface tunnel erosion, and fluvial erosion of both channel and bank) and mass movements of various types. Figure 1 illustrates these within a schematic representation of the physical landscape system.

Geomorphic landscape systems comprise a number of physical components, including their underlying lithology, soils, vegetation, water, fauna, and the landform assemblage that these form in their entirety. Depending on the scale of interest, various of these components may be considered as dynamic elements of the system, subject to variation, and others as external boundary conditions (Schumm and Lichty, 1965). In addition, landscape systems are subject to energy inputs and the application of force through a range of geomorphic processes. As with the physical components of the landscape, some energy inputs and processes may be regarded as a part of the system's boundary condition, while others act dynamically within the system.

Using the terminology of Chorley and Kennedy (1971) the flux of sediment through a landscape system can be conceived of as a process-response system. It thus involves a consideration of morphology, with landform as both a controlling factor (morphometry) and a product of the system's evolution. Further, a cascade of both matter and energy is implicit in the systematic definition of the landscape sediment flux. The sediment flux within this systematic context is thus envisaged as a series of sites – landforms – within the landscape in which sediment is stored for varying lengths of time (Figure 2). The long–term behaviour of the sediment flux is manifest as a change in landform (Figure 3).

However, reconstruction and modelling of the sediment flux over late Holocene time requires a more complex conceptualisation. Slope systems as represented in Figure 3 are not, in reality, closed; only in very rare circumstances will all of the sediment exported from slopes remain within such a simple source/transport/deposition system. This does not represent a great difficulty if one is only interested in a simple headwater slope. But the landscape is comprised

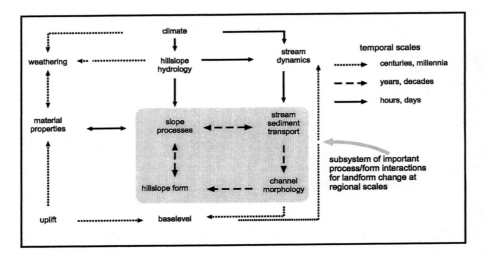

**Fig. 1.** Generalised flow chart representing the principal components and processes that are considered relevant to the long–term regional sediment flux.

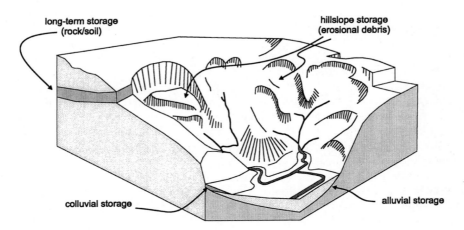

**Fig. 2.** Idealised landscape sketch, illustrating some of the storage components within regional landscape systems.

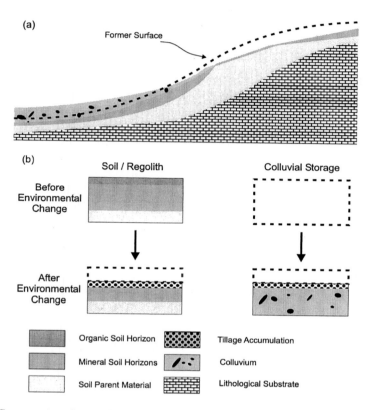

**Fig. 3.** Systematic sediment flux exemplified schematically with a hillslope system prior to anthropogenic change through deforestation, and as it exists after a period of agricultural land use. (a) Two dimensional representation of late Holocene hillslope change. Soil and regolith material has been eroded from upper slope areas, and redeposited as colluvium in lower parts of the landscape. (b) Schematic representation of the change of storage that this represents. The hillslope system is comprised of two sediment storage units. Before deforestation and introduction of agriculture, the soil/regolith storage unit is characterised as a maturely developed soil formed in in situ parent material, e.g. loess, while potential colluvial storage remains unoccupied. After some interval of time, and as a result of anthropogenic influences, the soil/regolith storage unit has been partly evacuated, and part of the colluvial storage unit is now occupied. Note the change in the soil/regolith unit; the natural A horizon and part of the *in situ* B horizon have been eroded, and both this and the colluvial storage units are characterised by the presence of a plough layer.

of many more elements than headwater slopes. Reconstructing the behaviour of the sediment flux within higher order landscapes must take account of sediment flux between systems such as this, incorporating all of the processes included in Figure 1.

These processes operate at different temporal and spatial scales, with a wide range of frequency/magnitude spectra. Crozier (1999:43) points out that the challenge facing contemporary geomorphologic research is "to isolate and predict the landform product of intersecting process regimes, interrelated but all working to their own frequency and magnitude agendas." Clearly this challenge applies not only to the prediction of landform products, but equally to the understanding of the landscape sediment flux.

Temporal and spatial scales are inextricably linked in geomorphic systems because of frequency/magnitude aspects of process behaviour, lag times in response to process occurrence, and especially because of complex response and the different reaction of landforms of different magnitude to identical perturbations (Schumm 1973; 1979). Walling (1983) has identified the issue of slope/channel coupling. Trimble (e.g. 1993; 1999) has demonstrated clearly how pulses of sediment can take decades to move through fluvial systems, and the widespread occurrence of these so-called "sediment slugs" has been summarised by Nicholas et al. (1995). This issue applies not only to fluvial systems, but also to the movement of sediment within slope systems. For example, Lang and Hönscheidt (1999) have developed a qualitative model for the flux of sediment that is specifically focused on colluvial redistribution of agricultural sediments on slopes.

Investigations in the Pleiser Hügelland, a region of loess-covered arable hill country near to Bonn in Germany, have indicated that the longer-term behaviour of the sediment flux is strongly influenced by two factors. The interaction between processes with different frequency/magnitude spectra on the one hand and, on the other, factors relating to the spatial configuration of landscape elements (Preston, 2001). In a landscape with a long history of agriculture, the considerable difference in the recurrence intervals of perturbations inducing low magnitude diffusive erosional processes and those that initiate high magnitude linear erosion has had consequences for the regional sediment flux, and thus for landform development (Preston, 2001). Similarly, Schmidt (2001) has demonstrated that when the landscape is considered over longer (Quaternary) time scales, mass movements in sensitive parts of the landscape must also be taken into account. These processes deliver – both directly and indirectly through subsequent reworking – large volumes of material to the regional sediment flux. Representation of sediment flux over long periods must therefore incorporate the contributions of a range of geomorphic processes that operate with very different frequency/magnitude spectra. Similarly, an increase in the spatial scale at which the sediment flux is represented introduces a range of further phenomena that must be recognised. Preston (2001) has shown that residence times of sediment in storage vary with different landscape locations, that the storage capacity of accumulation zones is a significant control on the sediment flux, and that the topological relationships between these various units is of crucial importance.

These findings support the argument of Lane and Richards (1997), that the spatial configuration of landscape components is an important influence on the flux of sediment.

The preceding discussion has attempted to demonstrate that representation of the sediment flux at the temporal and spatial scales inherent within the LU-CIFS programme requires that some important modelling issues be addressed. Importantly, it is the flux of sediment through the landscape that is of interest in the research agenda addressed by LUCIFS (Wasson, 1996), and not the behaviour of individual processes within limited spatial domains. It is argued, therefore, that models of sediment flux at large spatio-temporal scales must be able to integrate a range of processes operating with different frequency/magnitude spectra. Further, given the importance of storage capacity and connectivity for sediment redistribution, models must also be able to incorporate these factors.

## 3 Existing Models

This section presents a brief review of existing modelling approaches, and considers to what extent they enable representation of the sediment flux in the context of the points made in the previous section.

Modelling of sediment redistribution began as an attempt to address soil erosion problems. The most widely used approach is the Universal Soil Loss Equation (Wischmeier and Smith, 1978). This and its numerous refinements are often referred to as empirical models, i.e. they are based on observed statistically significant relationships. Input variables relate to rainfall, morphometry, soil erodibility, vegetation cover and land use factors and the response variable is a volume of material eroded or a net surface lowering. These are simple models of long–term average surface erosion at plot or small slope scales, and thus their most useful application lies in the identification of best agricultural practices for individual plots. Such an approach specifically does not model sediment redistribution through the landscape, and does not involve any physical representation of process behaviour.

Process–based models represent an increase in sophistication over empirical models. These are based on an understanding of the processes that underlie observed empirical relationships, and are often described as physically–based. Considerable knowledge has been gained in this respect, and at the field scale, at least, the physical basis of many erosional/depositional processes is well understood. More interesting for representation of sediment flux are spatially distributed process–based models i.e. those that include a spatial dimension, using a continuity equation satisfying elementary physical principles regarding conservation of mass and energy. Examples include Erosion3D (Schmidt, 1991; von Werner, 1995), WEPP (Nearing et al., 1989) and the EUROSEM model (Morgan et al., 1998). Discussion and assessment of these and similar models can be found in Boardman and Favis-Mortlock (1998a), and Siakeu and Oguchi (2000). They operate at sub-basin or basin level, subdividing space into discrete units (cells or modules), and modelling the movement of sediment between these units. Basin sediment outputs can be quantified, with applications for denudation and channel sedimentation studies. The more advanced are capable of elucidating within basin sediment redistribution, with resolution that is dependent on the scale of the internal unit. This approach is clearly a better representation of reality, and can provide further useful information for questions of applied geomorphology beyond the individual plot or slope.

These models specifically model sediment flux, with a more-or-less sound physical basis. Boardman and Favis-Mortlock (1998b) have suggested that they perform better over longer periods, with better results for continuous rather than event-based modelling. Indeed, in recent years such process models have been used for representing longer term rates of process behaviour, both with the objective of testing hypotheses about soil development (Favis-Mortlock et al., 1997) and of assessing possible future responses to climate and global change (e.g. Favis-Mortlock and Boardman, 1995; Favis-Mortlock and Savabi, 1996).

However, use of these models to represent regional sediment flux at a Holocene time scale is problematic. Extensive parameterisation is required, presenting not only potentially prohibitive computational demands but also the genuine challenge of obtaining sufficient, reliable and appropriate data. More importantly, many of the factors used in such models (e.g. those relating to vegetation properties and climatic drivers) are subject to change and variation through time. If these factors are treated as constant, i.e. part of model boundary conditions, then the period to which the model can be applied is determined by the frequency of variation in these factors. Therefore, although these are genuine sediment flux models, they can only be used easily at limited spatial and temporal scales.

The MEDALUS and MEDRUSH models (Kirkby 1998; 1999; Kirkby et al., 1998) represent another spatially distributed process–based approach. The MEDALUS models comprise a series of sub-routines dealing with slope hydrology, soil surface properties, vegetation growth and soil-atmosphere interactions. In combination, they calculate water and sediment fluxes for a point on a slope, and thus constitute a sophisticated process–based model of sediment entrainment through runoff processes. A spatial element is introduced by linking each point topologically to route sediment through a slope profile. MEDRUSH develops this concept further by linking MEDALUS point models in a thalweg sequence – a "flow strip" – which is considered representative of its basin. Each point within the sequence is considered representative of all areas within its elevation band. These "flow strips" link catchments, routing water/sediment through a drainage network. This approach is thus an attempt to regionalise process behaviour, and to model sediment redistribution between basins. Issues relating to the treatment of spatial scale remain, although this is to a large extent inevitable with regionalisation. Nevertheless, given the detail with which the processes of entrainment are treated, the approach is very data intensive, and is designed to cover periods of decades. Thus, while the spatial question is addressed, the applicability of this approach is temporally limited. There are ways to address this through parameter aggregation – as will be discussed later – but perhaps the biggest drawback to applying the MEDALUS/MEDRUSH models to Holocene time scales is that they are restricted in the range of processes that they represent. Specifically, mass movement processes, which contribute potentially very large volumes of material to the sediment flux over long periods, are not included.

More recently, a series of models have been developed that include multiple processes in attempting to represent landform development over Holocene (or longer) time scales. Prominent examples include the SIBERIA model (Willgoose et al., 1989; 1991a,b), CAESAR (Coulthard et al., 1996; 1997; 1999; Coulthard and Macklin, this volume) and CHILD (Tucker et al., 2001). As with the spatially distributed process–based models discussed above, these models incorporate water and sediment mass balances. They achieve their express objective of representing landform development by converting loss or gain of mass at a given point to a change in elevation. They can therefore be legitimately considered as being essentially sediment flux models. Coulthard (2001) provides a comparative review of these and other similar models; thus, although potentially the most promising approaches to modelling of the long–term sediment flux, they will only be briefly discussed here in terms of the broad strengths and weaknesses of the approach they represent.

Generally, the landform evolution models include a greater range of process descriptions, and the role of mass movement is recognised, albeit superficially in some cases. Not surprisingly, given that these are landform evolution models, they have generally been applied to greater spatial and temporal scales than the models described above. For example, Coulthard et al. (1996; 1997; 1999) used the CAESAR model to represent the development of an upland drainage basin over a period of some 9,000 years. The CHILD model (Tucker et al., 2001) is intended to be applied with a spatial resolution of $\sim 1\,km^2$, while the SIBERIA model (Willgoose et al. 1989; 1991a,b) has been applied to the development of river basins. An inevitable consequence of increasing scale has been a sacrifice of either spatial and/or temporal resolution. This means that processes are often represented more simply and the demand for high resolution input data has been reduced. Nevertheless, questions remain with regard to parameterisation, e.g. how are both internal properties and external drivers represented? Another issue is that of validation. Beven (1996) points out that this type of model may be able to predict plausible patterns, but has difficulty in predicting "real" landscapes. This is a genuine issue, but one that may apply to any approach dealing with long temporal scales; plausible patterns may well be the most realistic objective of long–term modelling. Similarly, the parameterisation issue can be addressed through aggregation techniques (as discussed below, and see also Favis-Mortlock et al., 1997). Nevertheless, it is thought that this model approach does not fully address the requirements of long–term regional scale modelling identified above, i.e. integration of processes with different frequency/magnitude spectra and incorporation of spatial configuration phenomena.

The model approaches that have been briefly reviewed here are summarised in Table 1. It is emphasised that this is a selected range of models, intended to be representative of the approaches available, and is by no means exhaustive.

**Table 1.** Summary of selected existing models that represent elements of the sediment flux. These are characterised in terms of the spatio-temporal scales they deal with, the processes that they include, and the degree of parameterisation they require. (References: (a) Schmidt 1991; von Werner 1995. (b) Nearing et al. 1989. (c) Morgan et al. 1998. (d) Kirkby 1998; 1999; Kirkby et al. 1998. (e) Willgoose et al. 1989; 1991a,b. (f) Coulthard et al. 1996; 1997; 1999. (g) Tucker et al. 2001.)

| Model | Space | | Time | | Process/es | Parameterisation |
|---|---|---|---|---|---|---|
| | Scale | Resolution | Scale | Resolution | | Demands |
| Erosion3D[a] | Hillslope | High | $10^1$ | Minute | Wash | Intermediate |
| WEPP[b] | Single Basin | Moderate | $10^1$–$10^2$ | Daily | Wash,rill | High |
| EUROSEM[c] | Single Basin | Moderate | Event | Minute | Wash,rill | High |
| MEDALUS[d] | Slope | * | $10^1$ | Minute | Wash | High |
| MEDRUSH[d] | Multiple Basin | Low | $10^1$ | Month | Fluvial transport | Low |
| SIBERIA[e] | Regional/ Continental | Low | $10^5$ | ** | Diffusive slope processes Fluvial Tectonic | Low |
| CAESAR[f] | Regional | High | $10^3$ | Daily | Wash Channel bank Channel bed Landsliding | High |
| CHILD[g] | Regional | Low | $10^5$ | >Event | Wash, rill Soil creep Channel bank Tectonics | Low |

* The MEDALUS model represents erosional processes for a single soil column.
** The temporal resolution of the SIBERIA model is arbitrary.

# 4  A proposed approach for modelling long–term regional sediment flux

It can be concluded that the existing approaches for modelling sediment redistribution summarised in Table 1 are generally focused on deterministic representation of individual processes. In this respect it is considered that they do not fully address the issues, referred to above, that are pertinent to modelling of the long–term regional sediment flux. These issues arise because of the shift in focus from individual processes to the behaviour of a sediment flux that is an integral of multiple processes. Based on this shift of focus, a conceptual approach for the modelling of long–term regional sediment fluxes is postulated here. The broad structure of the proposed modelling approach is illustrated in Figure 4, and outlined in the remainder of this section.

The concept of the process–response system is regarded as being scale-independent (see Crozier, 1999). It can be applied not only to simple discrete systems (e.g. the stability of a single slope and associated generation, transport and deposition of sediment), but also to complex, landscape scale sediment flux. The common link between morphological and cascading systems, and thus a defining characteristic of process–response systems, is generally in the form of an element of a morphological system that operates as a decision regulator within a cascade (Chorley and Kennedy, 1971). Within existing deterministic spatially distributed process–based models, morphology commonly plays the role of de-

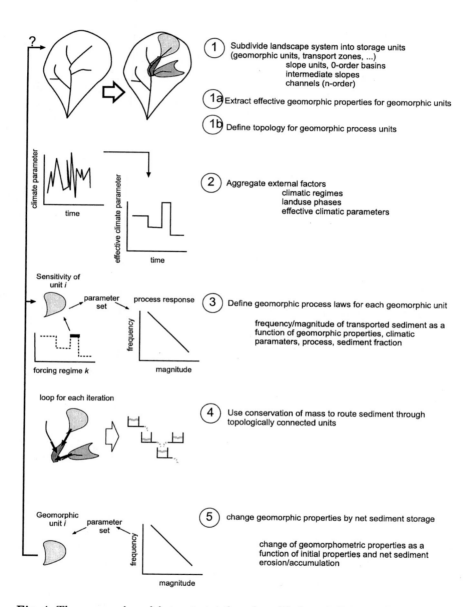

**Fig. 4.** The proposed model structure is based on (1) the subdivision of the landscape system into geomorphic storage units, (2) aggregation of external parameters (representing forcing processes and land use), (3) establishment of functional relationships between forcing processes, landscape sensitivity and sedimentary response, (4) routing sediment through topologically linked storage units, and (5) redefinition of morphology.

cision regulator, and there are many local scale morphometric properties that can be used for model parameterisation (e.g. Schmidt et al., 1998, Schmidt and Dikau, 1999). Similarly, the susceptibility of sediment to mobilisation and deposition can be seen as a function of morphometric landscape properties, and the morphology of storage units may fill the role of decision regulator in the context of a sediment flux model. In contrast to models of individual geomorphic processes, the morphometric properties that are likely to be relevant for the sediment flux will not be those with which the land surface is characterised. Rather, it will be those that determine potential storage capacity and transmissivity of sediment within and through the landscape. Storage capacity of individual storage units is important, but so too is the way in which this is influenced by coupling between those units, i.e. configuration and topology. Subdivision and characterisation of the landscape in these terms therefore forms the initial stage, and the framework, of the proposed approach.

Each unit within the landscape is conceived of as being capable of producing, storing and/or exporting a quantity of sediment within a given period. This approach can be seen as a simple tank model. It is analogous to the cell-based discretisation of existing spatially distributed process–based models, and sediment is routed through a series of storage components. However, this approach is intended for application at much longer time scales, and multiple processes must be integrated. Therefore an alternative means of expressing the relationship between forcing processes and the behaviour of the sediment flux is required. The approach proposed here is based on derivation of representative variables of external forcing processes through means of temporal aggregation, and the use of frequency/magnitude spectra to integrate the outputs of individual geomorphic processes.

At certain time scales climate can be considered as independent of the geomorphic system. Implicitly, however, meteorological forcing processes (i.e. precipitation, and streamflow as a first order derivative) are treated as dynamic internal components of the sediment flux system. Furthermore, land use is a significant control and modifier of process behaviour. Clearly, as was discussed with respect to existing modelling approaches, parameterisation of these factors for large temporal scales is problematic. In order to simplify the temporal variability of these factors, it is helpful to divide the continuous time scale into *climatic/land use regimes*, as exemplified by Favis-Mortlock et al. (1997) (Figure 5). These show comparative internal homogeneity in terms of the magnitude and behaviour of forcing processes, but statistically significant differences in representative values and variability. As a simple example, climate/land use regimes may be characterised with indices of rainfall erosivity and stream power. These factor values must be *effective* in terms of characterising the magnitude of response expected from each individual process within the given time period. The resolution of these aggregate parameter intervals will be dependent on the resolution with which climate proxies (e.g. Glaser et al., 1999) can be derived and the degree of information concerning historical land use that can be delivered.

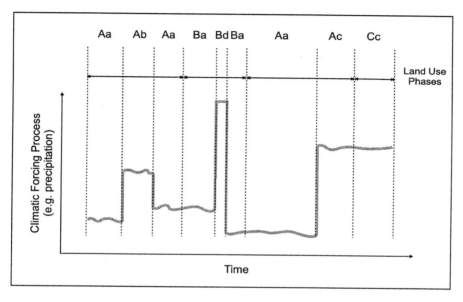

**Fig. 5.** Parameter aggregation: identification of periods with relative homogeneity of climate and land use. Representative values of external factors (e.g. precipitation, temperature, vegetation cover) are used to parameterise the proposed model for each of these homogeneous periods.

Magnitude of response is characterised with reference to functional relationships between forcing processes, sensitivity of the landscape and the sedimentary response. Each unit is characterised in terms of process domain, i.e. an indication of the relative significance of individual geomorphic processes. Establishment of process domains may be on the basis of observed precedence, or involve superposition of information layers describing the distribution of factors that influence process occurrence (vegetation, land use, lithology, rainfall regime, etc.). The unit sedimentary response is defined as the net product of all relevant erosional processes summed for the interval of interest, and is derived by reference to frequency/magnitude characteristics of net sediment redistribution for each geomorphic process. Two possibilities are envisaged for the derivation of appropriate frequency/magnitude spectra. Firstly, this may be achieved simply through empirical observation (e.g. Pearce, 1976; Gallart, 1995). Extensive databases exist already describing the frequency/magnitude spectra of various geomorphic processes, e.g. Boardman and Favis-Mortlock (1999) for wash erosion processes or Hovius et al. (1997) for landsliding. Where such information exists for all relevant processes, it can be combined within historical sediment budgets, quantifying the relative contributions of individual processes to regional sediment fluxes. Alternatively, representative frequency/magnitude spectra may be generated through simulation of individual process behaviour using stochastic parameterisation of existing process–based models (Schmidt, 2001). Importantly, frequency/magnitude spectra showing erosional/redistributional responses to perturbing events of various magnitudes are required, enabling responses to aggregated forcing process drivers.

As discussed above each unit is to be described by a set of *geomorphometric parameters*. These serve the purpose of defining storage capacity and, together with topology, of contributing to sediment export/receipt. This morphometric characterisation facilitates the development of empirical *form–process relationships* within the cascading system, where the *process* in question is that of *sediment redistribution*. Geomorphic/storage units are *linked topologically*, routing sediments through the whole landscape system. To complete the cascading system, functional *process–form relationships* are applied, e.g. through use of a simplified sedimentation law. Thus morphology and topology are redefined, establishing initial conditions for the next model iteration.

## 5   Discussion

This description of the model approach is brief, and is intended to introduce a concept for representation of the sediment flux at large spatio-temporal scales. It is clear, however, that various aspects of this approach require further consideration and research.

**Definition of geomorphic (storage) units** . While classification of landscape units on purely morphometric bases has been extensively dealt with in recent decades, semantic and scaling issues are not trivial. The proposed approach requires that units be defined relative to the system under consideration, which implicitly reflects the processes that are likely to be active, i.e. there is a *prima facie* semantic question. Although, scale may be considered arbitrary to an extent, there is an obvious implication with respect to the processes that are likely to occur within the system of interest. Thus, for example, both rotational landslides and micro-rills must be considered on a hillslope scale, but the latter can be discounted for modelling of the whole Rhine basin. An interesting approach that might be taken to the establishment of appropriate scale and resolution involves recognition of changing system states, and allowing variation in both spatial and temporal resolution. Depending on system condition with respect to perturbation, different processes may be expected to be active, and their relative importance will vary through time and space (see Crozier and Preston, 1999). For example, landsliding and the development of rills and gullies is to be expected on exposed hillslopes following deforestation; but, as hillslope supply diminishes, down-system fluvial processes may become more important for the overall flux of sediment. In the former case - immediately post-disturbance - the chosen scale must be able to reflect the importance of hillslope and headwater geomorphic units. As system evolution progresses, however, these may become less important, and a scale that recognises only channelised geomorphic units may suffice. Choice of scale might therefore be addressed using a preliminary spatial classification reflecting contemporary land surface condition (e.g. Preston, 1999) or time since perturbation (e.g. Crozier and Preston, 1999). Clearly, system evolution has implications for temporal scale and resolution as well, and these will be discussed below.

**Definition of representative and effective factors and establishment of phases within which these are relatively homogeneous.** This will inevitably be dependent on the resolution and the reliability with which historical data can be acquired and on the complexity of the external systems (climate system, human system). Thus the detail that can be achieved through aggregation will vary, and the sensitivity of the model to this will need to be evaluated.

**Derivation of frequency/magnitude spectra.** This is essentially an issue relating to the assemblage of empirical data sets and upscaling of small-scale process understanding. For low magnitude/high frequency processes this is not considered to be problematic. But for rare and catastrophic events, issues relating to the persistence within the landscape (Brunsden, 1993) of the effects of these processes will have to be addressed.

**Process/form coupling.** A crucial issue relates to the way in which geomorphically defined storage units are redefined after each model iteration. It is necessary firstly to determine the spatial distribution of sediment accumulation, and secondly to which degree it is possible to resolve this with respect to potential alteration in topology between units. Clearly, a new surface must be able to be defined in order to recognise the possibility of changing geomorphic configuration.

A related issue is that of boundary conditions. Within the (late) Holocene, lithology and geological processes, i.e. uplift/subsidence, are treated as being external to the system and part of its boundary conditions. Tectonic phenomena may indeed be active at the temporal scales envisaged, but it is reasonable to think that their effects will be uniform for the size of system for which this approach is envisaged, i.e. 3$^{rd}$ or 4$^{th}$ order drainage basins. While it is reasonable to treat tectonic phenomena as external to the system at these restricted spatial scales, this assumption is clearly not justified when modelling a system as extensive as the Rhine catchment. It is thus likely that a modular approach will be necessary, with some modules including a tectonic component and others disregarding this. This may be built into the model component addressing change in morphology; it is the effect of tectonics on relief that is important here (no attempt is made to address seismic triggering of mass movement).

**Establishment of appropriate iteration interval/s.** These will need to recognise order of magnitude differences in frequency/magnitude spectra, and also that different aspects of boundary conditions will be relevant at different spatio-temporal scales. It is envisaged that this approach should be hierarchical, i.e. applied to landscapes of different hierarchical level (or order). Thus, different processes will be important, different factors will determine boundary conditions, and therefore different iteration intervals will be appropriate dependent on the spatio-temporal character of the particular system in question. As discussed previously, system condition will also have some influence in determining appropriate model scale and resolution. In newly disturbed or transient systems, where high frequency, low magnitude processes occur, a high temporal resolution is likely to be necessary. In mature, stable or undisturbed systems, a lower temporal resolution may be sufficient. Clearly, the issues of spatial and temporal resolution are not independent, and the ability to vary both as necessary will be a key attribute of this approach.

## 6   Conclusion

The conceptual modelling approach that has been presented here addresses some of the principal geomorphic issues that are crucial for representing the sediment flux at large temporal and spatial scales. The focus must be regional, i.e. on higher order landscape systems, and it must be on the whole landscape scale process–response system, rather than on its individual process components. Over long temporal scales, it is difficult, if not impossible, to parameterise existing process models adequately. Thus, rather than representing individual processes, the proposed model focuses on the flux itself and incorporates a measure of the sediment contributed to that flux by each process on the basis of its frequency/magnitude spectrum. The approach therefore inherently deals with a multiplicity of processes, and attempts to address their interaction through recognition of their different frequency/magnitude regimes. Parameterisation of this model does not involve the factors that control the processes themselves - which are generally impossible to acquire for pre/historical periods. Rather, the model is parameterised with morphometric properties that influence the routing of sediment through the landscape. Most importantly, the proposed approach specifically emphasises geomorphic configuration over process representation. It is predicated on the thesis that representation of regional sediment fluxes and landform evolution requires models dealing with geomorphic configuration rather than detailed four-dimensional modelling of individual geomorphic processes. The topological relationships used in this approach are "real", rather than simply being model topology. Finally, the issue of parameter aggregation has been addressed. This is necessary for pre/historical periods for which data of high resolution cannot be obtained.

This is conceived as an overarching model that is designed for use in conjunction with other approaches. Specifically, within each geomorphic unit, existing process–based models – with greater or lesser resolution – can be used to generate inputs to the sediment flux, and perhaps further to enable analysis of the consequences for a given geomorphic unit of the sediment delivered to it. It is emphasised that this is not a model for estimating sediment yields from individual processes. Nor is it claimed that this approach will be capable of predicting the landform product of the sediment flux with any great accuracy. Rather, this approach is considered to offer a hierarchical modelling framework within which to address the questions posed by the LUCIFS programme.

## Acknowledgements

The ideas presented in this paper were developed during the authors' doctoral studies under the supervision of Professor Dr. Richard Dikau at Universität Bonn. We are grateful for that supervision and for an environment that stimulated discussion. We thank our colleagues in Bonn and in particular Dr. Andreas Lang and Dr. Sue Brooks. The constructive criticism of this manuscript offered by Dr. David Favis-Mortlock is greatly appreciated.

## References

Beven, K. (1996): Equifinality and uncertainty in geomorphological modelling. In: Rhoads, B.L., and Thorn, C.E. (eds.): The Scientific Nature of Geomorphology. Wiley, Chichester: 289–313.

Boardman, J., and Favis-Mortlock, D.(eds.): 1998. Modelling Soil Erosion by Water. NATO ASI Series 55. Springer, Berlin. 531 pp.

Boardman, J., and Favis-Mortlock, D. (1998): Modelling soil erosion by water: some conclusions. In: Boardman, J., and Favis-Mortlock, D. (eds.): Modelling Soil Erosion by Water. NATO ASI Series 55. Springer, Berlin: pp. 515–517.

Boardman, J., and Favis-Mortlock, D. (1999): Frequency–magnitude distributions for soil erosion, runoff, and rainfall – a comparative analysis. Zeitschrift für Geomorphologie Suppl. 115: 51–70.

Brunsden, D. (1993): The persistence of landforms. Zeitschrift für Geomorphologie Suppl. 93: 13–28.

Chorley, R.J., and Kennedy, B.A. (1971): Physical Geography – A Systems Approach. Prentice-Hall, London. 370 pp.

Coulthard, T.J. (2001): Landscape evolution models: a software review. Hydrological Processes 15: 15–173.

Coulthard, T.J., Kirkby, M.J., and Macklin, M.G. (1996): A cellular automaton landscape evolution model. In: Abrahart, R.J. (ed.): Proceedings of the 1st International Conference on GeoComputation. School of Geography, University of Leeds: 248–281.

Coulthard, T.J., Kirkby, M.J., and Macklin, M.G. (1997): Modelling hydraulic, sediment transport and slope processes, at a catchment scale, using a cellular automaton approach. In: Pascoe, R.T. (ed.): Proceedings of the 2nd Annual Conference: GeoComputation 97. Otago University, Dunedin: 309–318.

Coulthard, T.J., Kirkby, M.J., and Macklin, M.G. (1999): Modelling the impacts of Holocene environmental change in an upland river catchment, using a cellular automaton approach. In: Brown, A.G., and Quine, T.A. (eds.): Fluvial Processes and Environmental Change. John Wiley and Sons, New York: 31–46.

Coulthard, T.J., and Macklin, M.G. (*this volume*): Long–term and large scale high resolution catchment modelling: Innovations and challenges arising from the NERC Land Ocean Interaction Study (LOIS). In: Lang, A., Hennrich, K.P., and Dikau, R. (eds.): Long term hillslope and fluvial system modelling: Concepts and case studies from the Rhine river catchment, Lecture Notes in Earth Sciences, Springer, Heidelberg: 123–134.

Crozier, M.J., and Preston, N.J. (1999): Modelling changes in terrain resistance as a component of landform evolution in unstable hill country. In: Hergarten, S., and Neugebauer, H.J. (eds.): Process Modelling and Landform Evolution. Lecture Notes in Earth Sciences 78. Springer, Heidelberg: 267–284.

Crozier, M.J. (1999): The frequency and magnitude of geomorphic processes and landform behaviour. Zeitschrift für Geomorphologie Suppl. 115: 35–50.

Dikau, R. 1999. The need for field evidence in modelling landform evolution. In Hergarten, S., and Neugebauer, H.J. (eds.): Process Modelling and Landform Evolution. Lecture Notes in Earth Sciences 78. Springer, Berlin. pp. 3–12.

Favis-Mortlock, D., and Boardman, J. (1995): Nonlinear responses of soil erosion to climate change: a modelling study on the UK South Downs. Catena, 25: 365–387.

Favis-Mortlock, D., Boardman, J., and Bell, M. (1997): Modelling long–term anthropogenic erosion of a loess cover: South Downs, UK. The Holocene, 7(1): 79–89.

Favis-Mortlock, D., and Savabi, M.R. (1996): Shifts in rates and spatial distributions of soil erosion and deposition under climate change. In: Anderson, M.G., and Brooks, S.M. (eds.): Advances in Hillslope Processes. John Wiley and Sons: 529–560.

Gallart, F. (1995): The relative geomorphic work effected by four processes in rainstorms: a conceptual approach to magnitude and frequency. Catena, 25(1–4): 353–364.

Glaser, R., Brazdil, R., Pfister, C., Dobrovolny, P., Vallve, M.B., Bokwa, A., Camuffo, D., Kotyza, O., Limanowka, D., Racz, L., and Rodrigo, F.S. (1999): Seasonal temperature and precipitation fluctuations in selected parts of Europe during the sixteenth century. Climatic Change, 43: 169–200.

Hovius, N., Stark, C.P., and Allen, P.A. (1997): Sediment flux from a mountain belt derived by landslide mapping. Geology, 25(3): 231–234.

Kirkby, M.J. (1998): Modelling across scales: the MEDALUS family of models. In: Boardman, J., and Favis-Mortlock, D. (eds.): Modelling Soil Erosion by Water. Springer, Berlin: 161–173.

Kirkby, M.J. (1999) Landscape modelling at regional to continental scales. In: Hergarten, S., and Neugebauer, H.J. (eds.): Process Modelling and Landform Evolution. Lecture Notes in Earth Sciences. Springer, Berlin: 189–203.

Kirkby, M.J., Abrahart, R., McMahon, M.D., Shao, J., and Thornes, J.B. (1998): MEDALUS soil erosion models for global change. Geomorphology, 24: 35–49.

Lane, S.N., and Richards, K.S. (1997): Linking river channel form and process: time, space and causality revisited. Earth Surface Processes and Landforms, 22: 249–260.

Lang, A. (in press) A frequency analysis of phases of colluviation in the loess hills of south Germany. Catena.

Lang, A., and Hönscheidt, S. (1999): Age and source of colluvial sediments at Vaihingen–Enz, Germany. Catena, 38: 89–107.

Lang, A. and Nolte, S. (1999): The chronology of Holocene alluvial sediments from the Wetterau, Germany, provided by optical and $^{14}$C dating. The Holocene, 9(2), 207–214.

Lang, A., Bork, H.-R., Mäckel, R., Preston, N.J., and Dikau, R. (2000): Land use and climate impacts on fluvial systems during the period of agriculture – examples from the Rhine catchment. PAGES Newsletter, 8(3): 11–13.

Lang, A., Bork, H.-R., Mäckel, R., Preston, N.J., Wunderlich, J., and Dikau, R. (in press): Changes in sediment flux and storage within a fluvial system – some examples from the Rhine catchment. Hydrological Processes.

Morgan, R.P.C., Quinton, J.N., Smith, R.E., Govers, G., Poesen, J.W.A., Auerswald, K., Chisci, G., Torri, D., and Syczen, M.E. (1998): The European Soil Erosion Model (EUROSEM): a dynamic approach for predicting sediment transport from fields and small catchments. Earth Surface Processes and Landforms, 23(6): 527–544.

Nearing, M.A., Foster, G.R., Lane, L.J., and Finkner, S.C. (1989): A process–based soil erosion model for USDA–Water Erosion Prediction Project technology. Transactions of the American Society of Agricultural Engineers, 32: 1587–1593.

Nicholas, A.P., Ashworth, P.J., Kirkby, M.J., Macklin, M.G., and Murray, T. (1995): Sediment slugs: large scale fluctuations in fluvial sediment transport rates and storage volumes. Progress in Physical Geography, 19: 500–519.

Pearce, A.J. (1976): Magnitude and frequency of erosion by Hortonian overland flow. Journal of Geology, 84: 65–80.

Preston, N.J. (1999): Event–induced changes in landsurface condition – implications for subsequent slope stability. Zeitschrift für Geomorphologie Suppl., 115: 157–173.

Preston, N.J. (2001): Geomorphic response to environmental change: the imprint of deforestation and agricultural land use on the contemporary landscape of the Pleiser Hügelland, Bonn, Germany. unpub. PhD thesis, Universität Bonn. 129 pp.

Schmidt, Jo. (2001): The role of mass movements for slope evolution: conceptual approaches and model applications in the Bonn area. unpub. PhD, Universitaet Bonn.

Schmidt, Jo. and Dikau, R. (1999): Extracting geomorphometric attributes and objects from digital elevation models – semantics, methods, future needs. In: Dikau, R., and Saurer, H. (eds.): GIS for Earth Surface Systems – Analysis and Modelling of the Natural Environment. Schweizerbart'sche Verlagsbuchhandlung, Stuttgart: 153–173.

Schmidt, Jo., Merz, B., and Dikau, R. (1998): Morphological structure and hydrological process modelling. Zeitschrift für Geomorphologie Suppl., 112: 55–66.

Schmidt, Jü. (1991): A mathematical model to simulate rainfall erosion. Catena Suppl., 19: 101–109.

Schumm, S.A. (1973): Geomorphic thresholds and complex response of drainage systems. In: Morisawa, M. (eds.): Fluvial Geomorphology. Suny, Binghamton: 299–310.

Schumm, S.A. (1979): Geomorphic thresholds: the concept and its applications. Transactions of the Institute of British Geographers, 4: 485–515.

Schumm, S.A., and Lichty, R.W. (1965): Time, space, and causality in geomorphology. American Journal of Science, 263: 110–119.

Siakeu, J., and Oguchi, T. (2000): Soil erosion analysis and modelling: a review. Transactions of the Japanese Geomorphological Union, 21(4): 413–429.

Trimble, S.W. (1993): The distributed sediment budget model and watershed management in the Paleozoic Plateau of the upper midwestern United States. Physical Geography, 14: 285–303.

Trimble, S.W. (1999): Decreased rates of alluvial sediment storage in the Coon Creek basin, Wisconsin, 1975–93. Science, 285: 1244–1246.

Tucker, G.E., Lancaster, S.T., Gasparini, N.M., and Bras, R.L. (2001): The Channel–Hillslope Integrated Landscape Development Model (CHILD). In: Harmon, R.S., and Doe, W.W. (eds.): Landscape Erosion and Evolution Modeling. Kluwer Academic/Plenum Publishers, New York: 349–388.

von Werner, M. (1995): GIS–orientierte Methoden der digitalen Reliefanalyse zur Modellierung von Bodenerosion in kleinen Einzugsgebieten. unpublished PhD thesis, Freie Universität Berlin.

Walling, D.E. (1983): The sediment delivery problem. Journal of Hydrology, 65: 209–237.

Wasson, R.J. (1996): Land Use and Climate Impacts on Fluvial Systems during the Period of Agriculture – Research Project and Implementation. PAGES Workshop Report Series 96–2. 51 pp.

Willgoose, G.R., Bras, R.L., and Rodriguez–Iturbe, I. (1989): A Physically Based Channel Network and Catchment Evolution Model, TR322. Ralph M Parsons Laboratory, Massachusetts Institute of Technology, Cambridge, Massachusetts.

Willgoose, G., Bras, R.L., and Rodriguez–Iturbe, I. (1991a): A coupled channel network growth and hillslope evolution model. 1. Theory. Water Resources Research, 27(7): 1671–1684.

Willgoose, G., Bras, R.L., and Rodriguez–Iturbe, I. (1991b): A coupled channel network growth and hillslope evolution model. 2. Nondimensionalisation and applications. Water Resources Research, 27(7): 1685–1696.

Wischmeier, W.H., and Smith, D.D. (1978): Predicting Rainfall Erosion Losses. USDA Agricultural Research Service Handbook 537.

# Modelling the Geomorphic Response to Land Use Changes

Anton Van Rompaey, Gerard Govers, Gert Verstraeten, Kristof Van Oost, and
Jean Poesen

Laboratory for Experimental Geomorphology, K.U.Leuven, Redingenstraat 16, 3000
Leuven, Belgium

**Abstract.** In this paper the geomorphic response to land use changes
in central Belgium is modelled. A spatially distributed sediment delivery model (SEDEM) for the calculation of sediment delivery to rivers
is presented. An existing data set on sediment yield for 24 catchments
in central Belgium was used for calibration and validation of the model.
Next, a set of possible future land use scenarios was generated by means
of a probabilistic procedure that is based on historical map sequences. Finally, the predicted land use scenarios were used as an input for SEDEM.
The results point out that a relative reduction in arable land causes a far
more important relative reduction in soil erosion and sediment delivery
to the rivers.

## 1   Introduction

Soil erosion and sediment deposition processes are determined by four main
factors: (1) soil type, (2) climate, (3) topography and (4) land use. Although
all of these factors interact with human activity to a certain extent, it is obvious that land use is the most 'manageable' factor in terms of soil conservation
planning. The possible impact of land use changes on sediment fluxes is therefore an important topic on both the scientific and political agenda (e.g. the
Land–Use/Land–Cover Change (LUCC) programme, Turner et al., 1993). Many
studies have demonstrated how historical changes in land–cover have affected
sediment fluxes in drainage basins (e.g. Dearing, 1992; McIntyre, 1994; Trimble,
1999; Lang and Honscheidt, 1999; Evans et al., 2000). Suspension load measures
and dated sedimentation records have shown how deforestation and changes in
agricultural practices have influenced erosion and sediment transport processes.
These 'fieldwork'–based studies have provided a deeper insight in the interactions between land use changes and geomorphic processes. Such an approach,
however, is very labour–intensive which restricts its application to some case
studies. Moreover the method does not allow performing scenario analyses with
possible future land use patterns.

An alternative approach is to use numerical modelling to simulate the geomorphic response of a drainage basin. The whole modelling exercise consists
of two steps: (1) the simulation of possible future land use patterns and (2)

the modelling of geomorphic processes using the generated land use patterns as input data.

The application of geomorphic models at the scale of a drainage basin was hitherto rather problematic because of the low quality of the available input data. Recent studies have started to address this issue by proposing simplified long–term geomorphic models such as CSEP (Kirkby and Cox, 1995); SEMMED (de Jong et al., 1999), MIRSED (Brazier et al., 2001), and simplified derivations of the Universal Soil Loss Equation (e.g. Jäger, 1994; Van Dijk and Kwaad, 1998; Van Rompaey et al., 2000). The advantage of such simplified model structures is that they require a relatively limited amount of input data, which facilitates their application at larger areas. The potential of these relatively new simplified geomorphic models for geomorphic response studies has hitherto not fully been exploited.

In this paper a simple but distributed modelling structure (SEDEM) is proposed that allows the simulation of spatially explicit sediment delivery processes at the scale of a watershed. Next, a technique for the generation of future land use scenarios based on transformation probabilities is introduced. Finally, the geomorphic response of these possible changes is evaluated both on–site (in terms of soil loss) and off–site (in terms of sediment export).

## 2    Modelling sediment delivery at a regional scale

### 2.1    Lumped versus distributed sediment delivery models

Sediment delivery models are often based on an empirical lumped approach (e.g. Walling, 1983; Atkinson, 1995; Bazoffi et al., 1996; Verstraeten and Poesen, 2001). Sediment yield rates are estimated using average basin characteristics such as basin size, drainage density, mean slope, mean land cover, mean soil type etc. A typical lumped concept is the sediment delivery ratio (SDR). The sediment delivery ratio (SDR) is defined as the ratio of sediment delivered at the catchment outlet to gross erosion within the basin. The SDR represents the resultant of various processes involved between on–site soil erosion and downstream sediment yield. Several researchers (e.g. Klaghofer et al., 1992; Bazoffi et al., 1996; Vanoni, 1975; Roehl, 1962) developed sediment delivery equations from studies of basins located in particular regions. Sediment delivery ratios generally decrease with increasing basin size, indexed by the area or stream length (Ferro and Minacapilli, 1995). Vanoni (1975) suggested the use of the following power function:

$$SDR = k\,A^n (1) \tag{1}$$

Where: $k$, $n$: numerical constants (–); $A$: the basin area (km$^2$)

In general $n$ varies between -0.01 and -0.25. Lumped sediment delivery equations are frequently used for engineering applications such as the estimation of filling rates of retention ponds and the maintenance costs of hydro–electric power plants (USDI, 1987; Bazoffi et al., 1996).

The predictive ability of this kind of equation however, is limited to the particular regions they have been developed from. Inherent to this lumped approach is that it is not possible to take into account the spatial structure of land use and topography within the catchment on sediment delivery. This inherently limits their applicability to practical problems such as the evaluation of different land management strategies on sediment delivery.

In principle these problems can be solved by using a distributed model, whereby the eroded sediment is explicitly routed over the landscape towards the river system, allowing for sediment deposition when the transport capacity is exceeded. The dependence of the sediment delivery processes on local factors (sediment detachment, transport capacity and topography of the drainage basin) emphasises the need to use a spatially distributed model. However, the increasing complexity of physically based erosion/sediment deposition models such as EUROSEM (Morgan et al., 1998) or WEPP (Nearing et al. 1989) has a considerable impact on the data requirements. Often it is technically or financially impossible to gather high–precision data at the scale of a large catchment.

Pilotti and Bacchi (1997) proposed a distributed sediment delivery model that routes sediment according to a user defined topology. Sediment deposition occurs when the topographic gradient is lower than a given threshold. The accuracy of their approach however is not known since the model was not validated with measured sediment yield values. For individual fields and first order catchments soil erosion and sediment deposition models have been developed by Desmet and Govers (1995) and Van Oost et al. (2000). The predicted long– and medium–term soil redistribution patterns were validated using [137]Cs sampling. However, the necessary data to calibrate and validate such a model structure for large catchments are often not available.

In this paper SEDEM (SEdiment DElivery Model) is proposed that allows the simulation of sediment delivery processes at a catchment scale. The model was calibrated and validated using sediment export data from several catchments in central Belgium.

## 2.2 Modelling structure

The proposed model has three main components: (1) the assessment of a mean annual soil erosion rate for each grid cell, (2) the assessment of a mean annual transport capacity for each grid cell, and (3) a sediment routing algorithm that redistributes the produced sediment over the catchment taking into account the topology of the catchment and the spatial pattern of the transport capacity.

### Assessment of the mean annual soil erosion rate

The erosion component consists of an adapted version of the Revised Soil Erosion Equation (RUSLE, Renard et al., 1991). In order to adapt the RUSLE to a two–dimensional landscape, the upslope length was replaced by the unit contributing area, i.e. the upslope drainage area per unit of contour length. The latter procedure is explained in detail by Desmet and Govers (1996) and Van Oost et al.

(2000). Field observations indicate that this two–dimensional approach of the RUSLE not only accounts for interrill and rill erosion but also to some extent for ephemeral gully erosion as effects of flow convergence are explicitly accounted for (Desmet and Govers, 1997).

## Assessment of the mean annual transport capacity

For each grid cell a mean annual transport capacity $T_C$ (kg m$^{-1}$) was assessed. $T_C$ is the maximum mass of soil that can exit a grid cell per unit length of the downslope face of the square. Desmet and Govers (1995) and Van Oost et al. (2000) considered the mean annual transport capacity to be directly proportional to the potential rill (and ephemeral gully) erosion.

$$T_C = K_{TC}\, E_{PR} \qquad (2)$$

Where: $T_C$ = the transport capacity (kg m$^{-1}$ y$^{-1}$), $K_{TC}$ = the transport capacity coefficient (m), $E_{PR}$ = the potential for rill erosion (kg m$^{-2}$ y$^{-1}$).

The transport capacity coefficient $K_{TC}$ describes the proportionality between the potential for rill erosion and the transport capacity. It can be interpreted as the theoretical upslope distance that is needed to produce enough sediment to reach the transport capacity at the grid cell, assuming a uniform slope and discharge.

The potential for rill erosion ($E_{PR}$) is assessed by subtracting the potential interrill erosion ($E_{PIR}$) from the potential total erosion ($E_{PT}$).

$$E_{PR} = E_{PT} - E_{PIR} \qquad (3)$$

Where: $E_{PT}$ : potential total erosion (rill + interrill) (kg m$^{-2}$ y$^{-1}$), $E_{PR}$ : potential rill erosion (kg m$^{-2}$ y$^{-1}$), $E_{PIR}$ : potential interrill erosion (kg m$^{-2}$ y$^{-1}$).

A potential soil erosion rate is a theoretical erosion rate (kg m$^{-2}$ y$^{-1}$) assuming that the field is bare and no soil conservation measures are undertaken. The RUSLE facilitates the assessment of both actual soil erosion rates and potential soil erosion rates. In the latter case the $C$– and $P$–factor of equation 1 are set equal to one. The potential soil erosion rate ($E_{PT}$) can therefore be assessed as:

$$E_{PT} = R\,K\,L\,S \qquad (4)$$

Where: $E_{PT}$: the potential soil loss (rill and interrill) (kg m$^{-2}$ y$^{-1}$), $R$: rainfall erosivity factor (MJ mm m$^{-2}$ h$^{-1}$ y$^{-1}$), $K$: the soil erodibility factor (kg h MJ$^{-1}$ mm$^{-1}$), $L$: the slope length factor (-), $S$: the slope gradient factor.

According to McCool et al. (1989), the potential for interrill erosion ($E_{PIR}$) can be estimated as:

$$E_{PIR} = a\,R\,K_{IR}\,S_{IR} \qquad (5)$$

Where: $a$: a coefficient (-), $K_{IR}$: interrill soil erodibility factor (kg h MJ$^{-1}$ mm$^{-1}$), $S_{IR}$: interrill slope gradient factor (–).

Data are not available to estimate $K_{IR}$. Therefore it was assumed that $K_{IR}$ = $K$. The slope factor for interrill erosion was calculated using the equation proposed by Govers and Poesen (1988):

$$S_{IR} = 6.86\, S_g^{0.8} \tag{6}$$

Where: $S_g$ = slope gradient (m/m).

The coefficient $a$ in equation 5 was set at 0.6. The constant (6.86) in equation 6 was so chosen that the interrill erosion rate equals the rill erosion rate on a 0.06 slope at a distance of 65 m from the divide. These parameter values agree with observations of Govers and Poesen (1988). The exponent in equation 6 is equal to the value proposed by Foster (1982). The combination of equations 3, 4, 5 and 6 results in:

$$T_C = K_{TC}\,(R\,K\,L\,S - a\,R\,K_{IR}\,S_{IR}) = K_{TC}\,R\,K\,(L\,S - a\,S_{IR}) \tag{7}$$

$K_{TC}$-values for the different types of land use in the catchment have to be assessed by means of calibration.


**Sediment routing**

Once the mean annual erosion rate and the mean annual transport capacity are known at each grid cell, a routing algorithm can be used to transfer the eroded sediment from the source to the river network. Starting from the highest grid cells in the landscape the sediment is routed downslope. For each grid cell a continuous flowpath towards the river system is determined. The flowpath network needs to be topologically consistent, i.e. flowpaths may only end at the river channel and no circular flowpaths may exist (Desmet and Govers, 1995). Therefore a single–flow algorithm was used. In order to make sure that no flowpaths are interrupted, the DEM was corrected with a pit–removal algorithm (Clark Labs, 1999). The construction of flowpaths then starts by determining the outflow cell for the grid cell in the upper left corner of the DEM using the steepest descent criterion. This means that the lowest of the 8 neighbouring cells is selected as outflow cell. Next, the outflow cell of the former outflow cell is determined. This procedure is repeated until a stream channel, the border of the DEM or an existing flowpath is reached. Construction of flowpaths continues until an outflow cell is determined for each cell. When the flowpath reaches a river–cell all the sediment is delivered to the river. The construction of flowpaths is purely topographically determined. Since land use has not been taken in consideration flowpaths may cross several land use classes.

For each cell the amount of sediment input is added to the amount of soil erosion in that cell. If the sum of the sediment input and the local sediment production is lower than the transport capacity then all the sediment is routed further downslope. If this sum exceeds the transport capacity then sediment output from the cell is limited to the transport capacity. In the latter case, limited erosion will occur if the transport capacity exceeds the sediment input to the cell. If the transport capacity is lower than the sediment input, there will be sediment deposition.

Before the routing of sediment over the landscape can be carried out it is necessary to define the topological relations between the grid cells. The output of the model consists of a pixel–map representing the amount of erosion or sediment deposition at each pixel. Furthermore the amount of sediment that reaches the river channels is calculated. The sediment yield ($SY$ in $t\,y^{-1}$) can be expressed as an absolute value in $t\,y^{-1}$. An area specific sediment yield value ($SSY$ in $t\,ha^{-1}\,y^{-1}$) can be calculated when the absolute sediment yield value is divided by the catchment area.

### 2.3  Model calibration and validation

In order to calibrate and validate the model it was applied to 21 catchments, all situated in the hilly region of central Belgium (Figure 1). The region is situated at the transition zone from the coastal plains in the north and the lowland plateaus in the south. This transition zone is strongly incised by rivers draining to the north. The silt loam and sandy silt soils under agricultural land are very sensitive to soil erosion. The overall average soil erosion rate is c. $4.3\,t\,ha^{-1}\,y^{-1}$ in the study area (including forest and built–up area) (Van Rompaey et al., 2000). The average field size in the selected catchments is 2.3 ha. The land is especially vulnerable to erosion in spring and early summer when typical summer crops like sugar beet (*Beta vulgaris L.*) maize (*Zea mays L.*), potatoes (*Solanum tuberosum L.*) and chicory (*Cichorium intybus L.*) have a low vegetation cover (Vandaele and Poesen, 1995).

For all 21 catchments, ranging in size from 7 to 4873 ha the mean annual sediment yield was estimated using three different techniques, depending on the size of the catchment area and available data: (1) measurements of erosion and deposition volumes in catchments (2), measurement of suspended load in rivers and (3) measurement of sediment deposition volumes in retention ponds (Table 1) (Verstraeten and Poesen, 2001). The accuracy of sediment yield estimates was assessed at 20 % (one standard deviation) (Verstraeten and Poesen, in press). The sediment export values range from 55 to $3141\,t\,y^{-1}$.

Calibration of the RUSLE parameters for the Belgian Loess Belt had already been carried out by Bollinne (1982), Van Rompaey et al. (2000), and Van Oost et al. (2000). Therefore, the RUSLE–parameters in our model were taken from these studies. Only the transport capacity coefficients for arable land ($K_{TC\_A}$) and for non–eroding surfaces ($K_{TC\_NE}$), were calibrated. The use of a limited number of parameters for calibration implies that optimal values can unequivocally be determined.

The calibration procedure was carried out as follows. The dataset with 21 sediment yield observations were sorted according to the size of the catchment. Next, the sorted dataset was split in two parts according to the ranking number: the records with an even ranking number were used for calibration, the records with an uneven ranking number were used for validation.

**Table 1.** Measured and modelled sediment yield (the first 11 catchments were used for model calibration, the next 10 used for model validation). A detailed description of the catchments can be found elsewhere (Verstraeten and Poesen, 2001); $SY$ = Sediment Yield $(t\,ha^{-1})$; $SSY$ = Area Specific Sediment Yield $(t\,ha^{-1}\,y^{-1})$

| Catchment | Area (ha) | Obs. Method | Obs.SY SSY | Obs. SY | SEDEM SSY | SEDEM SY | Regress. SSY | Regress. SY |
|---|---|---|---|---|---|---|---|---|
| | | | **CALIBRATION DATASET** | | | | | |
| Sterrebeek | 7 | A | 55 | 7.9 | | | | |
| Hammeveld2 | 25 | A/B | 277 | 11.1 | | | | |
| Ganspoel | 117 | C | 562 | 4.8 | | | | |
| Zwedebeek | 176 | C | 282 | 1.6 | | | | |
| Holsbeek | 226 | A | 1053 | 4.7 | | | | |
| Nederaalbeek | 269 | C | 444 | 1.7 | | | | |
| Bellewaerdebeek | 1050 | A | 2384 | 2.3 | | | | |
| Kemmelbeek | 1138 | A | 3005 | 2.6 | | | | |
| Zouwbeek | 1362 | A | 1988 | 1.5 | | | | |
| Steenbeek | 1915 | A | 3141 | 1.6 | | | | |
| Stjansbeek | 4873 | A | 1852 | 0.4 | | | | |
| | | | **VALIDATION DATASET** | | | | | |
| Nerm | 20 | A | 411 | 20.6 | 319 | 15.9 | 259 | 7.1 |
| Hammeveld1 | 29 | B | 171 | 5.9 | 164 | 5.6 | 257 | 9.3 |
| Wolvengracht | 148 | A | 473 | 3.2 | 355 | 2.4 | 1046 | 4.2 |
| Nossegem | 206 | A | 723 | 3.5 | 121 | 0.6 | 493 | 3.2 |
| Kinderveld | 250 | B/C | 945 | 3.8 | 1004 | 4.0 | 977 | 2.1 |
| Ronebeek | 782 | A | 2420 | 3.1 | 1242 | 1.6 | 1379 | 3.1 |
| Munkbosbeek | 1102 | A | 1289 | 1.2 | 1129 | 1.0 | 1506 | 0.9 |
| Tererpenbeek | 1172 | C | 785 | 0.7 | 1113 | 0.9 | 1315 | 1.0 |
| Rooigembeek | 1394 | A | 3123 | 2.2 | 2745 | 2.0 | 1813 | 1.1 |
| Broenbeek | 2423 | A | 1381 | 0.6 | 841 | 0.3 | 1927 | 1.3 |

*(\*) Observation method: A = measurement of sediment deposition in retention ponds; B = measurement of erosion and sediment deposition volumes in catchments; C = suspended load measurements.*

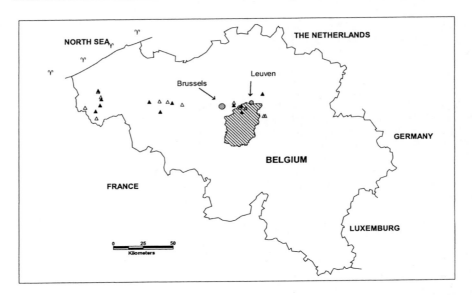

**Fig. 1.** Situation of the Dijle catchment (hatched area). Black triangles: catchments used for validation; white triangles: catchments used for validation (after Van Rompaey et al., 2001b)

For each catchment of the calibration dataset (n=11) the model was run with $K_{TC\_A}$ parameter values ranging from 0 to 150 and the $K_{TC\_NE}$ parameter values ranging from 0 to 60. For each combination of $K_{TC\_A}$ and $K_{TC\_NE}$ a sediment yield value was calculated for the 12 catchments. This allowed a comparison of the measured and predicted values for each parameter combination. The model efficiency coefficient ($ME$) as proposed by Nash and Sutcliffe (1970) was used as a measure of likelihood.

$$ME = 1 - \frac{\sum \left(Y_{obs} - Y_{pred}\right)^2}{\sum \left(Y_{obs} - Y_{mean}\right)^2} \tag{8}$$

Where: $ME$: the model efficiency, $Y_{obs}$: the observed value, $Y_{pred}$: the predicted value, $Y_{mean}$: the mean observed value. Values for $ME$ range from -∞ to 1. The closer $ME$ approximates 1, the better the model will predict individual values.

The results of these simulations are plotted in Figure **??** and **??**. The results show an optimal value for $K_{TC\_A}$ at 75 m and an optimal value for $K_{TC\_NE}$ at 42 m. The calibration curve of the $K_{TC\_NE}$ has a very flat top from which it may be concluded that the model when applied on rural areas with mainly arable land, is not very sensitive to this parameter. With the optimal parameter values a model efficiency coefficient of 0.77 is obtained.

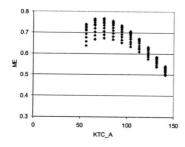

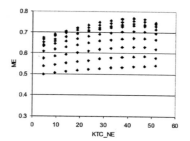

**Fig. 2.** Calibration of the transport capacity coefficient for arable land ($K_{TC\_A}$ in m); $ME$ = Model Efficiency.

**Fig. 3.** Calibration of the transport capacity coefficient for non–eroding surfaces ($K_{TC\_NE}$ in m); $ME$ = Model Efficiency.

Next the model was run for the catchments of the validation dataset (n = 10). Modelled versus observed values are plotted in Figure 4.

The model accuracy was estimated using the Relative Root Mean Square Error ($RRMSE$) as an error–measure (equation 9).

$$RRMSE = \frac{\sqrt{\frac{1}{n}\sum_{i=1}^{n}(O_i - P_i)^2}}{\frac{1}{n}\sum_{i=1}^{n}O_i} \tag{9}$$

Where : $O_i$ : observed sediment yield ($t\,y^{-1}$), $P_i$ : modelled sediment yield ($t\,y^{-1}$), n the number of catchments.

Using equation 9 resulted in a $RRMSE$ of 36 % from which may be concluded that 66 % of the model predictions have an error of less than 36 % ($1\,\sigma$). 95 % of the model predictions have an error of less than 72 % ($2\,\sigma$).

## 2.4   Model Application

For application at a larger spatial scale the Dijle catchment, south of Leuven, was selected (Figure 1). The basin–size is about 82,000 ha. The main topography is a plateau in which the Dijle and tributary rivers are incised. Height values in the area vary from 18 m above sea level in the north to 112 m in the south. The soils within the catchment are mainly loess–derived luvisols but on some places sandy outcrops exist. The land use in the catchment is mainly arable land (58 %) with some forest (13 %) and some pasture (5 %), mainly on valley floors. Built–up areas and infrastructure occupy 24 % of the land. Using the RUSLE, the average water erosion rate in the catchment, including the land under pasture and forest, was assessed at $4.4\,t\,ha^{-1}\,y^{-1}$. Transport capacities were calculated using the calibrated parameter–values for $K_{TC\_A}$ (75 m) and $K_{TC\_NE}$ (42 m).

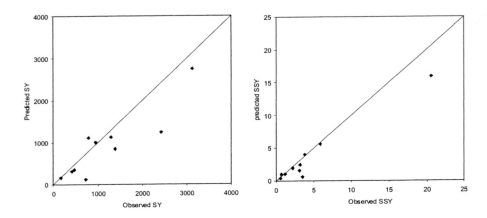

**Fig. 4.** Observed versus modelled sediment yield ($SY$ in $t\,y^{-1}$) and area–specific sediment yield ($SSY$ in $t\,ha^{-1}\,y^{-1}$) using SEDEM.

Routing the sediment through the catchment resulted in a map with net erosion and sediment deposition rates.

It is possible to identify the critical areas delivering most of the sediment to the river system. These areas do not necessarily have the highest erosion rates, as most of the eroded sediment is deposited within the catchment, before reaching the river system. The Dijle catchment was split up in 16 subcatchments (Figure 5). For each subcatchment an average erosion rate was calculated by taking the average of all the pixels of the erosion map. An average sediment export value was calculated by summation of the sediment input in all the river grid cells in the subcatchment. This value was divided by the area of the subcatchment. Soil conservation measures in areas with a high sediment export values will have the most effect on the suspended load in the river system.

Next, for each river grid cell the amount of sediment input ($SI$) ($t\,y^{-1}$) was calculated as well as the amount of upstream erosion ($UE$) ($t\,y^{-1}$). Next a distributed pattern of SDR–values was calculated using equation 10.

$$SDR_i = \frac{SI_i}{UE_i} \tag{10}$$

These SDR–values were plotted against the upstream catchment area (ha) of each cell (Figure 6). The SDR–values were put into area classes according to their upslope area. Figure 6 is similar to other SDR/upstream area plots but in this case all the dots belong to one catchment of which the SDR–value is gradually followed downstream. The variation in SDR is very high for pixels at the channel heads (low upstream area) and gradually decreases downstream.

The predicted SDR–value at the outlet of the catchment is ca. 18 %. This means that the model predicts a mean annual sediment export of 65,700 $t\,y^{-1}$

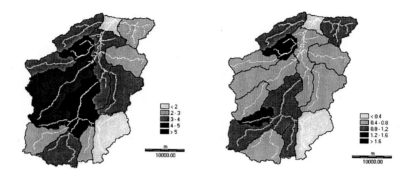

**Fig. 5.** Left: predicted mean annual soil erosion $(t\,ha^{-1}\,y^{-1})$ ; Right : predicted mean annual area specific sediment yield $(t\,ha^{-1}\,y^{-1})$ for subcatchments of the Dijle catchment. River channels are in white, borders of subcatchments in black (after Van Rompaey et al., 2001b)

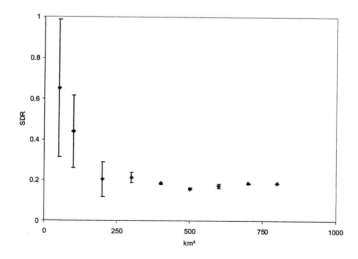

**Fig. 6.** Sediment delivery ratio's (SDR) versus upslope area for the river channels in the Dijle catchment. Pixels were divided in classes according to their upslope area. The graph shows the average SDR–value and the standard deviation of each class (after Van Rompaey et al., 2001b)

or $0.8\,\mathrm{t\,ha^{-1}\,y^{-1}}$. Since the percentage of bedload in the Dijle is negligible, long term suspended load data could give a good estimate of the annual sediment export. However suspended load measurements in the Dijle are available only for short time periods. Gilles and Lorent (1966) measured the suspended load during a 1–year period (1959–1960) taking one sample every 2 days and assessed the mean annual sediment yield at $0.3\,\mathrm{t\,ha^{-1}}$. However, the year during which the suspended load measurements were taken was rather dry. The mean annual precipitation in 1959 was only 594 mm whilst the long–term average annual precipitation is about 835 mm. Huybrechts et al. (1989) did similar suspended load measurements in 1985 and 1986 and assessed the mean annual sediment export at $0.7\,\mathrm{t\,ha^{-1}}$. Huybrechts et al. (1989) took only one sample every month, so that their values are almost certainly underestimated. The most recent suspended load measurements were done by Steegen (pers. comm.) taking two samples a day with an automatic ISCO sediment sampler. After 14 months of measurement during a very wet period the mean annual sediment yield was assessed at $2.1\,\mathrm{t\,ha^{-1}\,y^{-1}}$ Thus, all reported values are in the same range as the prediction of our model. However, for a real validation of the model results, suspended load measurements over a longer time period are necessary. Data for a validation of the distributed pattern of erosion and sedimentation values within the Dijle catchment are not available. Nevertheless, this simulation indicates that the modelling approach developed here may also be used for larger river catchments. It should be kept in mind, though, that the proposed model only allows the prediction of sediment production and delivery from water erosion on hillslopes. Non–rill processes like land slides or debris flows are not included. In larger river systems, fluvial processes like floodplain sediment deposition and river bed and bank erosion may affect or even dominate the sediment balance. The model structure presented here may still be used then to predict the sediment input into the river system by water erosion, but as a result data on river sediment loads can then no longer be used for model calibration or validation.

# 3   Modelling land use changes

## 3.1   Conditional probabilities and the simulation of future land use scenarios

Land use change models that may be coupled with these geomorphic models must address the location issue: 'Where will what land use take place?' At present, economical, sociological or political driving forces induce the majority of the land use transformations (e.g. Turner et. al, 1990; Stoorvogel, 1995; Veldkamp and Fresco, 1996; De Koning et al, 1998; Thornton and Jones, 1998; Van Rompaey et al., 2001a). The spatial pattern of these transformations is, however, determined by both socio–economic and physical parameters. Given the difficulty of specifying the future evolution of such driving forces, most of the existing land use change models aim at the generation of possible land use scenarios that can be used as an input for geomorphic response models. In this paper a method is proposed that allows the generation of future land use patterns based on land use changes in the past.

If –for a test site– a set of recorded land use transformations is available it is possible to distinguish two groups: (1) a group of land units with a transformed land use (e.g. from arable land to forest or vice versa) and (2) a group of land units with a stable land use over the considered time period. Next, variables describing the difference between both groups have to be found. Such explaining variables may be bio–physical (e.g. soil properties, slope gradient) or socio–economic (e.g. accessibility, distance from market?). By means of statistical testing it is possible to evaluate for each available variable whether or not the two groups have significantly different properties. If a variable is significant, it can be used to calculate transformation probabilities. A transformation probability is the probability that a field will be converted from arable land into fallow land. For a single variable (e.g. soil texture) a transformation probability can be estimated as the relative frequency of transformed parcels in that category (e.g. a soil texture class). More then one parcel characteristic being known, the overall transition probability can then be calculated using the theorem of Bayes:

$$P\left(A_i \cap B_i \cap ... \cap N_i\right) = \frac{P\left(trans \mid A_i\right) \cdot P\left(trans \mid B_i\right) \cdot ... \cdot P\left(trans \mid N_i\right)}{P\left(trans\right)^{n-1}}$$

$$(11)$$

Where:

$A_i$ : the i$^{th}$ class of variable $A$ (e.g. the class $\ll$ 10-15 % $\gg$ of Slope)
$B_i$ : the i$^{th}$ class of variable $b$ (e.g. the class $\ll$ Sand $\gg$ of Texture)

The average transition probability $P(trans)$ depends on the selected scenario. In the case of a 5 % transformation scenario $P(trans)$ is 5 %. Equation 11 can be extrapolated over a larger area in order to calculate for each land unit a transformation probability.

Based on the calculated transformation probabilities, stochastic simulations of future land use transformations may be carried out as follows: A land unit in the catchment is selected at random. Next a random number between 0 and 1 is chosen. This number is then compared with the transformation probability of the selected parcel. If the random number is less than the transformation probability then the land unit is accepted for a land use transformation; if not the original land use of the land unit is kept. This procedure is re–iterated until the desired simulated percentage of forest is reached. This way of parcel selection implies that parcels with a higher transition probability have a higher probability of being selected for forestation. Nevertheless, it should be kept in mind this selection procedure assumes that the transition rules didn't change over time.

The use of the method described above will be illustrated by means of two example applications in the Dijle catchment in central Belgium.

## 3.2   Example 1: Forestation and deforestation probabilities derived from historical maps

Historical land use patterns were reconstructed for 4 test–sites in the north of the Dijle catchment. Four large scale topographic maps were available: the Ferraris maps (1774), the Vandermaelen maps (1840) and the maps of the National Geographic Institute of Belgium (1930, 1990) (Table 2).

**Table 2.** Topographic maps used in this study.

|  | Ferraris map | Vandermaelen map | Nat. Geographic Institute of Belgium (NGI) | Nat. Geographic Institute of Belgium (NGI) |
|---|---|---|---|---|
| Date of publication | 1744 | 1840 | 1930 | 1990 |
| Original scale | 1:11,500 | 1:20,000 | 1:20,000 | 1:10,000 |
| Survey | ground | ground | ground | photogrammetric |
| Planimetric precision after resampling | $25\,m/km$ | $18\,m/km$ | $5\,m/km$ | $5\,m/km$ |

A detailed description of these map series can be found elsewhere (Depuydt, 1991). On each map 4 different land use classes were distinguished: arable land, forest, pasture and built–up area. The delineated polygons were digitised and resampled by means of a local polynomial resampling method (Clark Labs, 1999). The vector–files for these 4 georeferenced land use maps were converted to raster–maps with a grid–size of $5\,m * 5\,m$ (Figure 7).

The land use maps were then compared with the land use maps of the next time–period by means of an overlay operation. This resulted in 3 land use transformation maps: 1774–1840, 1840–1930 and 1930–1990. Each map shows the land use transformations from one time–period into another (e.g. from arable land to forest, from arable land to pasture, from arable land to arable land, etc) (Figure 8).

Since the land use maps have 4 different land use classes, the land use transformation maps can have a maximum of 16 classes (stable land use included). However, it appeared that a lot of possible transformations did not occur in the test sites. 83 % of the transformations in the study area were 'forest – arable land' or 'arable land – forest'. Only these transformations were considered in this study. The test sites were subdivided in smaller units with an average size of 2.5 hectares, which is more or less the actual average field size on arable land.

For each of the generated parcels, the mean slope was calculated by means of a digital elevation model (DEM). In order to find out to what degree the slope gradient of a field determines the probability of a land use conversion, the parcels were split into two parts for each time period separately: parcels with a stable land use during the considered time period, and parcels with a change in

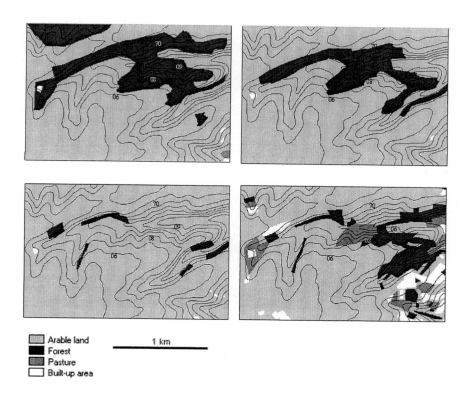

**Fig. 7.** Land use pattern in 4 different time periods (test site Ganspoel). Top left: Ferraris–map (1774), top right: Vandermaelen map (1840), bottom left: NGI (1930), bottom right: NGI (1990). Contour interval $= 10\,m$ (after Van Rompaey et al., 2002).

**Fig. 8.** Land use transformation in the Ganspoel test site in 1930–1990 (after Van Rompaey et al., 2002).

land use. If slope does not play a role in the conversion of land use, then there should be no difference between the mean slope of the converted parcels and the mean slope of the non–converted fields. This hypothesis was tested via a Student t–test. The results of the t–tests are listed in Table 3.

**Table 3.** Statistical analysis of the mean slope gradient (slope gradients are expressed in %).

| Transition | Total arable land before transition | | Arable land after transition | | Forest after transition | | T–value | Prob. T |
|---|---|---|---|---|---|---|---|---|
| | number of parcels | mean slope | number of parcels | mean slope | number of parcels | mean slope | | |
| 1774–1840 | 323 | 4.6 | 319 | 4.5 | 4 | 9.8 | 3.32 | 0.0010 |
| 1840–1930 | 344 | 4.8 | 335 | 4.7 | 9 | 8.7 | 4.46 | 0.0002 |
| 1930–1990 | 323 | 4.7 | 284 | 4.7 | 22 | 8.1 | 4.46 | 0.0006 |
| Transition | Total forest before transition | | Forest after transition | | Arable land after transition | | T–value | Prob. T |
| | number of parcels | mean slope | number of parcels | mean slope | number of parcels | mean slope | | |
| 1774–1840 | 53 | 10.2 | 11 | 12.5 | 24 | 9.2 | 2.82 | 0.008 |
| 1840–1930 | 15 | 11.8 | 4 | 16.6 | 11 | 10.1 | 2.43 | 0.028 |
| 1930–1990 | 12 | 16.8 | 10 | 17.7 | 2 | 12.6 | 1.13 | 0.270 |

The mean slope gradient of parcels with an arable land – forest conversion (upper part of Table 3) is significantly higher than the slope of non–converted parcels with arable land. For each of the three transformations periods the probability of the T–values is lower than 0.1. On the other hand, the mean slope gradient of the parcels with a forest–arable land conversion (lower part of Table 3) is significantly lower than the non converted forest parcels. This trend, however, is only significant in two of the three transformation periods because of the small amount of parcels in this category. The results of the statistical analysis show that slope plays a significant role when land use is converted from arable land to forest and vice versa. For conversions from arable land to forest, parcels with steep slope gradients are preferred, while conversion from forest to arable land are on parcels with a relatively low slope gradient.

A soil texture map was compiled on the basis of soil cores of 1 meter depth with a density of 1–2 drillings per ha (Ameryckx et al., 1985). Soil textures were classified using the Belgian texture classification (Figure 9). Seven soil texture classes are distinguished: clayey silt loam (A), sandy silt loam (L), light sandy loam (P), loam sand (S), sand (Z), clay (E) and heavy clay (U). Map purity estimates indicate that c. 35 % of the observations were classified in a textural class just above or below the correct textural class (Van Meirvenne, 1998).

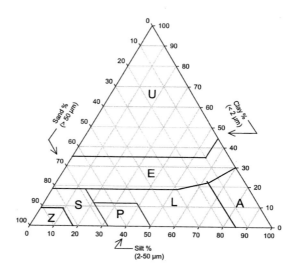

**Fig. 9.** Belgian textural classes: A = clayey silt loam, L = sandy silt loam, P = light sandy loam, S = loam sand, Z = sand, E = clay, U = heavy clay.

Soil drainage class was also mapped. Two classes were distinguished: well–drained and poorly drained soils. Via an overlay of the field–file and the soil texture and soil drainage map, it was possible to extract for each field the soil texture class and the soil drainage class. The original texture classes were reduced to 4 main texture classes: loamy soils (A), sandy loamy and sandy soils (L + P + S + Z) and clayey soils (E + U). This was done because some of the original texture classes contained very few members which would make analysis at this level of detail unreliable.

A $\chi^2$–test was used for testing whether soil properties of fallow fields are not significantly different from the non–fallow fields. A $\chi^2$–test compares the expected frequency distribution over classes with the observed distribution and calculates the probability of the two distributions being equal. The expected distribution of the converted parcels over the different texture classes was calculated as the product of the total distribution of all the parcels in the database and the relative frequency of the converted parcels in the total amount of parcels. The same procedure was carried out for soil drainage classes. The expected distribution over three main drainage classes (well drained and poorly drained) was compared with the observed distribution. A comparison of the observed frequency distribution over the different soil classes texture classes shows that parcels with a sandy or a clay soil texture are more likely to be converted to forest than fields with a loamy soil texture. Forests on soils with a loamy texture are more likely to be converted to arable land (Table 4).

**Table 4.** Expected versus observed distribution over soil texture classes.

| Soil texture | Transition 1774 – 1840 | | | |
| | Arable land to forest | | Forest to arable land | |
| | Expected # | Observed # | Expected # | Observed # |
| --- | --- | --- | --- | --- |
| Clay | 0 | 0 | 1 | 0 |
| Loam | 4 | 3 | 17 | 22 |
| Sand | 0 | 1 | 6 | 2 |
| **Transition 1840 – 1930** | | | | |
| | Arable land to forest | | Forest to arable land | |
| | Expected # | Observed # | Expected # | Observed # |
| Clay | 0 | 1 | 0 | 0 |
| Loam | 8 | 3 | 10 | 11 |
| Sand | 1 | 5 | 1 | 0 |
| **Transition 1930 – 1990** | | | | |
| | Arable land to forest | | Forest to arable land | |
| | Expected # | Observed # | Expected # | Observed # |
| Clay | 1 | 1 | 0 | 0 |
| Loam | 19 | 17 | 2 | 1 |
| Sand | 2 | 4 | 0 | 1 |
| **All transition** | | | | |
| | Arable land to forest | | Forest to arable land | |
| | Expected # | Observed # | Expected # | Observed # |
| Clay | 1 | 2 | 1 | 0 |
| Loam | 31 | 23 | 29 | 34 |
| Sand | 3 | 10 | 7 | 3 |
| | $\chi^2 : 19.9$ | Prob $> \chi^2 : 4E - 5$ | $\chi^2 : 4.1$ | Prob $> \chi^2 : 0.12$ |

This trend is observed in each of the three transformation periods. Because of the low frequencies in certain soil texture classes, a $\chi^2$–test was carried out using the distribution of all land use transformations together. The results confirm the observed trend at significance levels of 0.004 % and 12 %.

The results of the $\chi^2$–test point out that parcels on poorly drained soils are more likely to be converted to forest than parcels on well drained soils. Forests on dry soils are more likely to be converted to arable land (Table 5). This trend is observed in each of the three transformation periods at significance levels of 12 % and 14 %.

Slope gradient, soil texture and soil drainage were therefore taken into account for the simulation of land use change patterns. Equation 11 was used to calculate forestation and deforestation transformation probabilities (Tables 6 and 7).

**Table 5.** Expected versus observed distribution over soil drainage classes.

| Soil drainage | **Transition 1774 – 1840** | | | |
| | Arable land to forest | | Forest to arable land | |
| | Expected # | Observed # | Expected # | Observed # |
| Dry | 3 | 1 | 17 | 20 |
| Wet | 1 | 3 | 7 | 4 |
| | **Transition 1840 – 1930** | | | |
| | Arable land to forest | | Forest to arable land | |
| | Expected # | Observed # | Expected # | Observed # |
| Dry | 7 | 6 | 10 | 10 |
| Wet | 2 | 3 | 1 | 1 |
| | **Transition 1930 – 1990** | | | |
| | Arable land to forest | | Forest to arable land | |
| | Expected # | Observed # | Expected # | Observed # |
| Dry | 16 | 13 | 0 | 1 |
| Wet | 6 | 9 | 2 | 1 |
| | **All transition** | | | |
| | Arable land to forest | | Forest to arable land | |
| | Expected # | Observed # | Expected # | Observed # |
| Dry | 26 | 22 | 27 | 31 |
| Wet | 9 | 13 | 10 | 6 |
| | $\chi^2$ : 2.39 | Prob $> \chi^2$ : 0.12 | $\chi^2$ : 2.19 | Prob $> \chi^2$ : 0.14 |

**Table 6.** Forestation probabilities (all transition periods in the 4 test sites are taken into account).

| | | Total arable land land (# parcels) | Not converted | Converted to forest | Trans. prob. (%) |
| --- | --- | --- | --- | --- | --- |
| Slope class | < 5% | 569 | 561 | 8 | 0.014 |
| | 5 − −10% | 297 | 290 | 7 | 0.023 |
| | 10 − −15% | 79 | 71 | 8 | 0.101 |
| | > 15% | 45 | 33 | 12 | 0.266 |
| Texture class | Clay | 8 | 6 | 2 | 0.250 |
| | Loam | 950 | 927 | 23 | 0.024 |
| | Sand | 32 | 22 | 10 | 0.312 |
| Drainage class | Dry | 759 | 737 | 22 | 0.029 |
| | Wet | 231 | 2128 | 13 | 0.056 |

**Table 7.** Deforestation probabilities (all transition periods in the 4 test sites are taken into account).

|          |            | Total forest (# parcels) | Not converted | Converted to arable land | Trans. prob. (*) |
|----------|------------|:-----:|:-----:|:-----:|:-----:|
| Slope    | < 5%       | 12 | 2  | 10 | 0.66 |
| class    | 5 − −10%   | 18 | 6  | 12 | 0.61 |
|          | 10 − −15%  | 15 | 8  | 7  | 0.46 |
|          | > 15%      | 17 | 9  | 8  | 0.47 |
| Texture  | Clay       | 2  | 2  | 0  | 0.00 |
| class    | Loam       | 50 | 16 | 34 | 0.68 |
|          | Sand       | 10 | 7  | 3  | 0.30 |
| Drainage | Dry        | 52 | 21 | 31 | 0.59 |
| class    | Wet        | 10 | 4  | 6  | 0.60 |

(*) *Single factor transition probability.*

By means of the simulation procedure, described in section 3.1, forestation and deforestation scenarios were simulated. Starting from the present situation in the catchment (58 % arable land, 13 % forest), 10 land use scenarios were generated.

## 3.3   Example 2: Conditional probabilities derived from present–day land use changes

Since the reform of the EU–CAP in 1992, the support system for farmers in member states of the EU has thoroughly changed in order to control the level of agricultural production and to secure the income of the farmers. In the new support arrangements, a farmer has to set aside a minimum percentage of land. Since 1992 the minimum set–aside percentage has varied between 5 and 10 % in Belgium (Table 8).

Protection by vegetation of the fallow fields with selected fallow–species is obligatory: however the fallow species are specific for each EU–member state (Sibbensen, 1997). As the potential for runoff and soil erosion is very much affected by land cover, these CAP–regulations have a strong impact on the sediment fluxes in agricultural regions. A permanent vegetation cover protects the soil from direct raindrop impact, crusting and sealing which reduces the amount of surface runoff till almost zero. Moreover, fallow crops will trap sediment coming from other non–protected fields resulting in a decrease of sediment delivery to the river channels. The use of spatially distributed soil erosion and sediment delivery models allows assessing the impact of EU–CAP scenarios. This however requires the generation of realistic land use transition patterns based on the decision criteria of farmers.

**Table 8.** Minimum set–aside percentages in Belgium since the reform of the Common Agriculture Policy of the European Union (EU–CAP).

| Year | Minimum set–aside precentage (%) |
|------|-----------------------------------|
| 1992 | 15 |
| 1993 | 15 |
| 1994 | 15 |
| 1995 | 12 |
| 1996 | 10 |
| 1997 | 5 |
| 1998 | 5 |
| 1999 | 10 |
| 2000 | 10 |

Through an inquiry (summer 1997) among 31 farmers in the catchment a database was collected with 974 fields (Table 9).

**Table 9.** Parcels of the 31 questioned farmers (in summer 1997).

| Category[a] | Frequency | Relative frequency (%) |
|-------------|-----------|------------------------|
| Category I | 57 | 5.9 |
| Category II | 195 | 20.0 |
| Category III | 722 | 74.1 |
| Total | 974 | 100 |

[a] *Cat. I: Fallow at the moment of the inquiry*
*Cat. II: No fallow at the moment of the inquiry but a possibly next year*
*Cat. III: No fallow and none considered in the future*

The farmers were asked to classify their fields in 3 categories: (1) fallow at the moment of the inquiry, (2) no fallow at the moment of the inquiry but possible in the near future, and (3) no fallow and none considered in the near future (Figure 10).

For each field three attributes were calculated: (1) the mean slope using a digital elevation model with a resolution of 20 m, (2) the soil texture class and (3) the soil drainage class. Next statistical tests were carried out to test whether or not certain field characteristics are significantly different for fallow and non–fallow fields. A comparison between expected and observed frequency distribution for the three field attributes showed that farmers significantly prefer flat, well–drained fields on loamy soils for cultivation. Fields with steep slopes on sandy or clay soil and with a bad drainage have a higher probability to be selected for set–aside (Table 10).

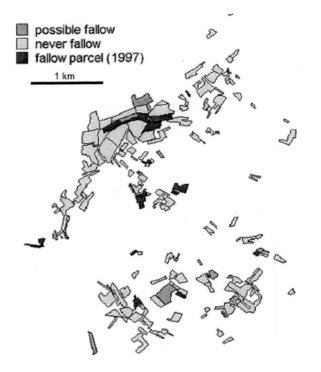

**Fig. 10.** Fallow and non–fallow parcels in 1997 (1 farm in the Dijle catchment).

**Table 10.** Expected versus observed frequency of fallow parcels.

|  | Expected frequency of fallow parcels | Observed frequency of fallow parcels |
|---|---|---|
| **Slope gradient class** | | |
| low ($< 5\%$) | 174 | 134 |
| moderate ($5–10\%$) | 13 | 16 |
| high ($10–15\%$) | 8 | 11 |
| very high ($> 15\%$) | 6 | 13 |
| **Soil texture class** | | |
| loam | 188 | 168 |
| samdy loam | 31 | 43 |
| sand | 25 | 28 |
| clay | 8 | 13 |
| **Soil drainage class** | | |
| Dry | 206 | 179 |
| Medium | 38 | 52 |
| Wet | 8 | 21 |

Via equation 11 the transition probabilities for every field in the catchment were assessed. Next by means of a random simulation (Van Rompaey et al., 2001a), set–aside volumes of 5 %, 10 %, 15 % and 20 % were simulated taking into account the slope and soil conditions of the fields (Figure 11).

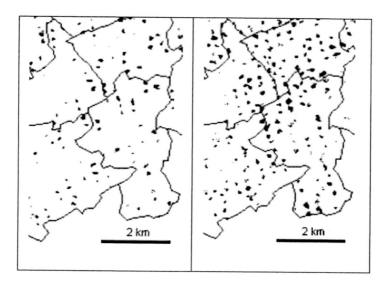

**Fig. 11.** Simulated set–aside patterns for some municipalities in the Dijle catchment (left 5 % under fallow, right 10 %).

# 4   Coupling land use and sediment delivery models

The simulated land use patterns were used as input for the RUSLE and SEDEM, as described in section 2. For each scenario the 'geomorphic response' (i.e. the mean annual soil erosion rate and the total sediment yield) was calculated. In this modelling exercise, we assume step–wise changes between different land use change patterns. The transitionary responses themselves, that may be higher than the new steady state, are not modelled.

An increase of the percentage of arable land (Figure 12) results in a faster than linear increase of the mean soil erosion rate in the catchment because the slope gradients of the newly deforested areas are systematically higher than the slope gradients on the original arable land.

Afforestation (or fallow), on the other hand, results in a faster than linear decrease of the mean annual soil erosion rate in the catchment because parcels with steep slopes are more likely to be converted into forest. A decrease of the

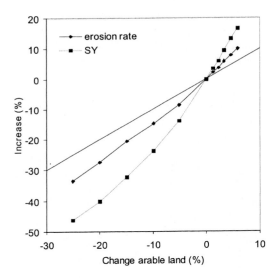

**Fig. 12.** Impact of land use changes on the soil erosion rate and the sediment yield (SY) (after Van Rompaey et al., 2002).

percentage arable land in the catchment by for example 5 % results in a 8.5 % lowering of the mean annual soil erosion rate.

The detected trend is even more significant when looking at the predicted sediment yield values ($SY$) because of a double effect : (1) there are less sediment sources and (2) there are more sediment sinks. If the percentage of arable land in the catchment increases by 5 %, there is 14.5 % more sediment delivered in the river channels. A 5 % decrease of the percentage of arable land results in a 13.5 % lowering of the sediment yield. The latter observation implies that the sediment delivery ratio is increasing as the percentage arable land increases.

Geomorphic response curves are important with respect to the design of a policy aiming at a protection of villages against flooding and muddy floods. These phenomena are often caused by a reduced retention capacity of channels and reservoirs because of sediment deposition. Obliged fallow and reforestation programmes may need to be taken into consideration because of their far reaching positive effects on sediment delivery and related problems.

## 5   Conclusions

In this paper a suite of models is applied to assess the impacts of land use change on geomorphic processes. Firstly, it is shown that a rather simple model with only elementary process descriptions (SEDEM) can be used to predict the sediment delivery by surface runoff from hillslopes to rivers in small catchments with an acceptable accuracy. Secondly, a technique is presented that applies historical map sequences to assist predicting possible future land use patterns. Finally, SEDEM is used to quantify the geomorphic process activity in terms of soil erosion and sediment delivery rates for each of the possible land use scenarios.

The results point out that the geomorphic response to land use changes is non–linear: a small change in percentage arable land results in relatively large changes in erosion risk and sediment delivery. In the Dijle catchment sediment delivery appears to be more sensitive to land use change than soil erosion due to the sediment trapping effect of forest or fallow vegetation. In catchments with a less pronounced topography, however, much of the eroded sediment is deposited even on fields with arable land. In catchments with such a morphology a partial reforestation may therefore be less effective with respect to the reduction of the sediment yield.

The final results of the modelling exercise presented here cannot be directly validated, as long–term sediment transport records for the Dijle catchment are not available. However, the SEDEM model was calibrated and validated in the same physiographical region, using data from various small catchments with important variations in land use and morphology. Its application to scenario analysis is therefore based on the ergodicity principle, whereby is assumed that observations in time can be substituted by observations in space. In the future a direct evaluation may be possible if more data on the long term evolution of colluvial and alluvial deposits are collected and land use evolution is more extensively and thoroughly documented.

# References

Ameryckx, J., Verheye, W., and Vermeire, R. (1985): Bodemkunde. J. Ameryckx, Gent, Belgium.

Atkinson, E. (1995): Methods for assessing sediment delivery in river systems. Hydrological Sciences 40(2): 273–280.

Bazoffi, P., Baldasarre, G., and Vasca, S. (1996): Validation of the PISA2 model for the automatic assessment of reservoir sedimentation. In: Albertson, M. (ed.) Proceedings of the International Confererence on Reservoir Sedimentation. Colorado State University: 519–528.

Bollinne, A., 1982. Etude et Prvison de l'érosion des sols limoneux cultivés en moyenne Belgique. Unpublished PhD Thesis, Université de Liège, Belgium.

Brazier, R.E., Rowan, J.S., Anthony, S.G., and Quinn, P.F. (2001): "MIRSED" towards an MIR approach to modelling hillslope soil erosion at the national scale. Catena 42(1): 59–79 .

Clark Labs (1999): Idrisi Guide to GIS and Image Processing. Clark University, Worcester, USA.

Dearing, J.A. (1992): Sediment yields and sources in a Welsh upland lake–catchment during the past 800 years. Earth Surface Processes and Landforms 17: 1–22.

Depuydt, F. (1991): Five centuries of cartography in Flanders. De Aardrijkskunde 4: 403–423.

Desmet, P.J.J., and Govers, G. (1995): GIS–based simulation of erosion and deposition patterns in an agricultural landscape: a comparison of model results with soil map information. Catena 25: 389–401.

Desmet, P.J.J., and Govers, G. (1996) A GIS procedure for automatically calculating the USLE LS factor on topographically complex landscape units. Journal of Soil and Water Conservation 51: 427–433.

Desmet, P.J.J., and Govers, G. (1997): Two–dimensional modelling of the within field variation of rill and gully geometry and location related to topography. Catena 29 : 283–306.

De Koning, G.H.J., Veldkamp, A., and Fresco, L.O. (1998): Land use in Ecuador: a statistical analysis at different aggregation levels. Agriculture, Ecosystems and Environment 70: 231–247.

De Jong, S.M., Paracchini, M.L., Bertolo, F., Folving, S., Megier, J., and De Roo, A.P.J. (1999): Regional assessment of soil erosion using the distributed model SEMMED and remotely sensed data. Catena 37(3–4): 291–308.

Evans, J.K., Gottgens, J.F., Gill, W.M., and Mackey, S.D. (2000): Sediment yields controlled by intrabasinal storage and sediment conveyance over the interval 1842–1994 : Chagrin River, Northeast Ohio, USA. Journal of Soil and Water Conservation 55(3): 264–270.

Ferro, V., Minacapilli, M. (1995): Sediment delivery processes at basin scale. Hydrological Sciences Journal 40(6): 703–717.

Foster, G.R., 1982. Modeling the erosion process. In: Haan, C.T., Johnson, H.P., and Brakensiek, D.L. (eds): Hydrologic Modeling of Small Watersheds. 533 pp. ASAE, St. Joseph, USA.

Gilles, M., and Lorent, J. (1966): Debiet en lading van de Dijle. Acta Geographica Lovaniensa 4: 48–56.

Govers, G., and Poesen, J. (1988): Assessment of the interrill and rill contributions to total soil loss from an upland field plot. Geomorphology 1: 343–354.

Huybrechts, W., Verbeelen, D., and Van der Beken, A. (1989): Meting van het sedimenttransport in de Dijle te Korbeek–Dijle. Water 45: 55–59.

Jäger, S. (1994): Modelling regional soil erosion susceptibility using the USLE and GIS. In: Rickson (ed.): Conserving Soil resources: European Perspectives. CAB International, Wallingford: 161–177.

Kirkby, M.J., and Cox, N.J. (1995): A climatic index for soil erosion potential (CSEP) including seasonal and vegetation factors. Catena 25(1–4): 333–352.

Klaghofer, E., Summer, W., and Villeneuve, J.P. (1992): Some remarks on the determination of the sediment delivery ratio. IAHS Publ. No. 209: 113–118.

Lambin, E.F., Rounsevell, M.D.A., and Geist, H.J. (2000): Are agricultural land–use models able to predict changes in land–use intensity? Agriculture, ecosystems and environment 82: 321–331.

McCool, D.K., Foster, G.R., Mutchler, C.K., and Meyer, L.D. (1989): Revised Slope Length Factor for the Universal Soil Loss Equation. Transactions of the ASAE 32(5): 1571–1576.

McIntyre, S.C., (1994): Reservoir sedimentation rates linked to long–term changes in agricultural land use. Water Resources Bulletin 29(3): 487–495.

Nash, J.E., and Sutcliffe, J.V. (1970): River flow forecasting through conceptual models. Part I. A discussion of principles. Journal of Hydrology 10: 282–290.

Morgan, R.P.C., Quinton, J.N, Smith, R.E., Govers, G., Poesen, J.W.A., Auerswald, K. Chisci, G., Torri, D., and Styczen, M.E. (1998): The European soil erosion model (EUROSEM): A dynamic approach for predicting sediment transport from fields and small catchments. Earth Surface Processes and Landforms 23: 527–544.

Nearing, M.A., Foster, G.R. Lane, L.J., and Finkner, S.C. (1989): A process–based soil erosion model for USDA–Water Erosion Prediction Project Technology. Transactions of the ASAE 32: 1587–1593.

Pilotti, M., and Bacchi, B. (1997): Distributed evaluation of the contribution of soil erosion to the sediment yield from a watershed. Earth Surface Processes and Landforms 22: 1239–1251.

Renard, K.G., Foster, G.R., Weesies, G.A., and Porter, J.P. (1991): RUSLE Revised Soil Loss Equation. Journal of soil and Water Conservation 46(1): 30–33.

Roehl, J.E. (1962): Sediment source areas, delivery ratios and influencing morphological factors. IAHS Publ. No. 59: 202–213

Sibbesen, E. (1997): Set–aside and land–use regulations with relation to surface runoff in Finland, Denmark and Scotland, Belgium, France and Spain. SP–report No. 14 – Danish Institute of Agricultural Science.

Stoorvogel, J.J. (1995): Geographical information systems as a tool to explore land charateristics and land use, with reference to Costa Rica, PhD thesis, Wageningen Agricultural Univerity, The Netherlands.

Thornton, P.K., and Jones, P.G. (1998): A conceptual approach to dynamic agricultural land use modelling. Agricultural Systems 57(4): 505–521.

Trimble, S.W. (1999): Decreased rates of alluvial sediment storage in the Coon Creek basin, Wisconsin, 1975–93. Science 285: 1244–1246.

Turner II, B.L., Clark, W.C., Kates, R.W., Richards, J.F., Matthews, J.T., and Meyer, W.B. (1990): The earth as transformed by human action. Cambridge University Press, UK.

Turner II, B.L., R.H. Moss, and Skole, D.L. (1993): Relating Land Use and Global Land–Cover Change: A Proposal for an IGBP–HDP: Core Project. IGBP Report 24 and HDP Report 5. International Geosphere–Biosphere Programme and the Human Dimensions of Global Environmental Change Programme, Stockholm.

USDI (1987): Design of small dams. A Water Resources Technical Publication. United States Department of the Interior, Bureau of Reclamation, 3th ed, Denver, Colorado, USA.

Vandaele, K., and Poesen, J. (1995): Spatial and temporal patterns of soil erosion rates in an agricultural catchment, central Belgium. Catena 25: 213–226.

Van Dijk, P.M., and Kwaad, F.J.P.M. (1998): Modelling suspended sediment supply to the river Rhine drainage network; a methodological study. In: Summer, W., Klaghofer, E., and Zhang, W. (eds): Modelling soil erosion, sediment transport and closely related processes. IAHS Publication 249: 165–176.

Vanoni, V.A. (1975): Sedimentation Engineering. ASCE Manuals and Reports on Engineering Practices, no. 54.

Van Oost, K., Govers, G., and Desmet, P.J.J. (2000): Evaluating the effects of landscape structure on soil erosion by water and tillage. Landscape Ecology 15(6): 579–591.

Van Rompaey, A., and Govers, G. (2000): Modelling of soil erosion on a regional scale : A case study in Flanders. In: Gabriels, D., Schiettecatte, W. (Eds). Proceedings of the Meeting Contactgroup Erosion. I.C.E. Special Report 2: 1–12.

Van Rompaey, A., Govers, G., Van Hecke, E., and Jacobs, K. (2001a): The impacts of land use policy on the soil erosion risk: a case study in central Belgium, Agriculture. Ecosystems and Environment 83: 83–94.

Van Rompaey, A., Verstraeten, G., Govers, G., Van Oost, K., and Poesen, J. (2001b): Modelling mean annual sediment yield using a distributed approach. Earth Surface Processes and Landforms 26: 1221–1236.

Van Rompaey, A., Govers, G., and Puttemans, C. (in press): Modelling land use changes and their impact on soil erosion and sediment supply to rivers. Earth Surface Processes and Landforms.

Verstraeten, G., and Poesen, J. (2001): Factors controlling sediment yield from small intensively cultivated catchments in a temperate humid climate. Geomorphology 40: 123–144.

Verstraeten, G., and Poesen, J. (in press): Using sediment deposits in small ponds to quantify sediment yield from small catchments: possibilities and limitations. Earth Surface Processes and Landforms.

Veldkamp, A., and Fresco, L.O. (1996): CLUE: a conceptual model to study the conversion of Land Use and its Effects. Ecological Modelling 85: 253–270.

Walling, D.E. (1983): The sediment delivery problem. Journal of Hydrology 65: 209–237.

# Modelling Water and Tillage Erosion using Spatially Distributed Models

Kristof Van Oost, Gerard Govers, Wouter Van Muysen, and Jeroen Nachtergaele

Fysische en Regionale Geografie, K.U. Leuven, Redingenstraat 16, 3000 Leuven, Belgium

**Abstract.** Soil erosion models are valuable tools for understanding sedimentary records. In this paper, the potential use of a topography–based model (WaTEM) for simulating long–term soil erosion and its effect of soil properties is discussed. Long–term (derived from profile truncation) and medium–term (derived from $^{137}$Cs measurements) erosion patterns are compared with simulated patterns of water and tillage erosion. Results showed that WaTEM is able to describe to reproduce the observed spatial pattern of long–term water erosion reasonably well. However, the $^{137}$Cs data indicated that a major change in erosion and sedimentation patterns has occurred over the last decades: the dominance of water erosion over a time scale of several thousands of years explains the spatial pattern of soil truncation. On the other hand, the $^{137}$Cs data indicate that the present–day pattern of soil erosion is dominated by tillage. WaTEM is also used to assess the effect of changes in landscape structure on soil erosion. It was shown that, when shifting focus from the field to the catchment scale, the way we represent space in distributed models affects the model performance at least as dramatically as the physical description of the process. Finally, a model application whereby WaTEM is linked with a mass–balance model, showed that simulating the effects of soil erosion on the redistribution of soil properties is an important issue when trying to link surface processes and sedimentary records.

## 1 Introduction

At present, a variety of soil erosion models exist focussing on different spatial (from point to catchment) and temporal scales (from event–based to long–term) and each of them having different degrees of complexity. When using soil erosion models as a tool for understanding sedimentary records or predicting sediment export to rivers, the model user should be aware of the possibilities and limitations of soil erosion models. In this paper, the potential of topography–based models for simulating long–term soil erosion and its effect on soil properties are discussed.

First, a topography–based model called WaTEM, is described and used to model spatial patterns and rates of water and tillage erosion. Model results are compared with long–term soil erosion rates (derived from soil profile truncation data) and medium–term erosion rates (derived from $^{137}$Cs measurements) in order to illustrate the potential use of this type of model.

Secondly, attention is given towards modelling soil erosion on the catchment scale. When shifting focus from individual fields to landscapes or catchments, the structure of the landscape, i.e. the spatial organisation of different land units and the connectivity, becomes a key factor in the soil erosion process. The major effect of spatial implementation, i.e. the way in which space is described in distributed models, is illustrated by simulating soil erosion rates in catchments where significant changes in landscape structure have occurred.

Until now, most erosion models focus on the prediction of soil erosion rates and sediment export to rivers. Little attention is given towards the effect of soil erosion on the spatial and temporal variability of soil properties. In this paper, a model is used to illustrate this variability and implications for studying soil profiles and sedimentary records are discussed.

## 2   WaTEM description

In this section, the process equations for water and tillage erosion used in Wa-TEM (Water and Tillage Erosion Model, Van Oost et al. (2000a)) are described in detail.

### 2.1   Water erosion component

The water erosion component of WaTEM consists of an adapted version of the Revised Soil Loss Equation (RUSLE). Use of the RUSLE is preferred, as parameter values for this equation are readily available for many areas. Despite the widespread acceptance of the RUSLE, it has two important disadvantages: (i) the RUSLE is limited in predicting soil loss on complex topographies and two–dimensional landscapes (ii) and the RUSLE does not predict where the eroded material will be deposited. In order to adapt the RUSLE to a two–dimensional landscape, the upslope length is replaced by the unit contributing area, i.e. the upslope drainage area per unit of contour length. The latter can be calculated using a procedure proposed by Desmet and Govers (1996):

$$L_{i,j} = \frac{\left(A_{i,j-in} + D^2\right)^{m+1} - A_{i,j-in}^{m+1}}{D^{m+2} x_{i,j}^m \left(22.13\right)^m} \quad (1)$$

Where: $L_{i,j}$ = the slope length factor for the grid cell with coordinates $(i,j)$ (-), $A_{i,j-in}$= the contributing area at the inlet of a grid cell with coordinates $(i,j)$ (m$^2$), $D$ = the grid cell side length (m), $x_{i,j} = sin\alpha_{i,j} + cos\alpha_{i,j}$, $\alpha_{i,j}$ = aspect direction for the grid cell with coordinates $(i,j)$ and $m$ = slope length exponent (-).

Field observations indicate that this two–dimensional approach of the RUSLE not only accounts for interrill and rill erosion but also to some extent for (ephemeral) gully erosion as effects of flow convergence are explicitly accounted for (Desmet and Govers, 1997). The sedimentation routine used in WaTEM is based on the work of Govers et al. (1993). The model describes the potential for rill erosion as a power function of slope length and gradient and the potential for interrill erosion as a power function of slope gradient. The potential annual erosion rate is considered equal to the sum of the potential for rill and interrill erosion unless the local transporting capacity is exceeded. The transport capacity on a given slope segment was considered to be directly proportional to the potential for rill erosion. If the sediment inflow exceeds the transport capacity, deposition occurs, so that the amount of material leaving a point equals the transport capacity. This approach may be adapted to be used within the RUSLE. The potential annual soil loss as predicted by the RUSLE can be used to replace the sum of interrill and rill erosion.

$$E_{PT} = RKLSCP \tag{2}$$

Where: $E_{PT}$ = potential mean annual soil loss as predicted by the RUSLE ($\mathrm{kg\,m^{-2}}$), $R$ is the rainfall erosivity factor, $K$ is the soil erodibility factor, $LS$ is the topographical factor, $C$ is the crop factor and $P$ is the management factor.

The transporting capacity can then be estimated as:

$$T_c = K_{TC}E_{PR} \tag{3}$$

Where $T_c$ = the transport capacity ($\mathrm{kg\,m^{-1}}$), $k_{TC}$ = transport capacity coefficient (m).

$EPR$ is the potential annual rill erosion:

$$E_{PR} = E_{PT} - E_{PIR} \tag{4}$$

According to McCool et al. (1989), potential interrill erosion ($E_{PIR}$) can be estimated as:

$$E_{PIR} = aRK_{ir}S_{ir}C_{ir} \tag{5}$$

Where: $a$ = a coefficient, $K_{ir}$ = interrill soil erodibility factor, $S_{ir}$ = interrill slope steepness factor, $C_{ir}$ = interrill soil management factor. Data are not available to estimate separately $K_{ir}$ or $C_{ir}$ and therefore we assumed that

$$K_{ir} = K \text{ and } C_{ir} = C$$

The slope factor was calculated as:

$$S_{ir} = 5.0S^{0.8} \tag{6}$$

The constant (5.0) was chosen so that the interrill erosion rate equals the rill erosion rate on a 0.06 slope at a distance of 65 m from the divide. This agrees with the data of Govers and Poesen (1988), collected on loam and sand loam soils. If no such data is available, the transport capacity can be approximated by assuming that $T_c$ is proportional to $E_{PT}$.

Together, these modifications of the RUSLE allow predicting water erosion and sedimentation rates and patterns for complex, two–dimensional landscapes if values for $R,K,C,P$ and $K_{TC}$ are available.

These equations are implemented in a grid–based model. Topographical attributes (slope and contributing area) are derived from a Digital Elevation Model. The eroded sediment is routed over the DEM using a multiple flow algorithm starting from the highest grid cell in the landscape towards the outlet of the catchment. Equations 1 to 6 are used to calculate the local transport capacity, sediment transfer and erosion/deposition rate.

## 2.2   Tillage erosion component

Tillage erosion results from variations in tillage translocation over a landscape. When soil is moved in the upslope direction by a tillage implement, the translocation distance will be smaller than when soil is moved downslope. Consequently, a net soil displacement results if soil is moved alternatively in the up– and downslope direction. The net flux due to tillage translocation on a hillslope of infinitesimal length and unit width is proportional to the local slope gradient (Govers et al. 1994):

$$Q_{s,t} = k_{til}S = -k_{til}\frac{dh}{dx} \tag{7}$$

Where $Q_{s,t}$ = the net downslope flux due to tillage translocation, $k_{til}$ = tillage transport coefficient, $S$ = the local slope gradient, $h$ = height at a given point of the hillslope and $x$ = distance in horizontal direction.

The local erosion or deposition rate ($E_t$) can then be calculated as:

$$E_t = \rho_b\frac{dh}{dt} = -\frac{dQ_{s,t}}{dx} = k_{til}\frac{d^2h}{dx^2} \tag{8}$$

However, soil tillage is an anisotropic process that strongly interacts with relief morphology (De Alba, in press; Van Muysen et al., in press a). Simulating tillage erosion in a two–dimensional context requires a more complex approach whereby tillage direction is explicitly accounted for:

$$Q_{s,long} = d_{long}D\rho_b \text{ and } Q_{s,lat} = d_{lat}\rho_b \tag{9}$$

Where $Q_{s,long}$ and $Q_{s,lat}$ are the soil fluxes in the tillage and lateral direction respectively, $d_{long}$ and $d_{lat}$ are the average displacement distances of the soil in the tillage and lateral direction and $D$ is the tillage depth.

The relationship between $d_{long}$, $d_{lat}$ and slope gradient can be derived from tillage erosion experiments (eg Van Muysen et al., 1999, in Press a).

Equation 8 implies that tillage erosion is controlled by the change in slope gradient, not by the slope gradient itself, so that erosion takes place on convexities while soil accumulation occurs in concavities. These equations describing tillage erosion are also implemented in a grid–based model. Topographical attributes (slope in tillage and lateral direction) are derived from a Digital Elevation Model. Soil fluxes are calculated for each grid cell and distributed towards neighbouring grid cells in both directions. The local erosion or deposition rate is then calculated using equation 8.

## 2.3 Evaluation of Simulated spatial patterns

The spatial patterns of soil redistribution are shown in Figure 1. A comparison of the slopes and the spatial patterns of water erosion rates points out that the areas with the steepest slopes have the highest erosion rates. This is simply because the slope gradient is the major control on the $LS$–value. Deposition occurs when the transport capacity is exceeded. The location of the depositional area and the amount of deposited sediment is controlled by the $k_{TC}$ value. The pattern of tillage erosion rates is quite different: high erosion rates occur on the convexities but also on the downslope side of field borders. The combination of water and tillage erosion leads to extreme high rates of erosion on steep, convex slope segments. Here, water and tillage erosion are maximized. On the other hand, deposition by tillage counteracts water erosion in thalwegs.

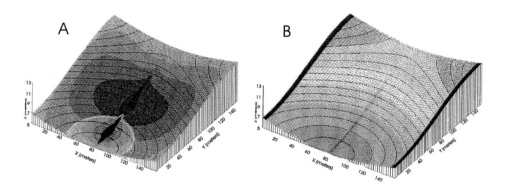

**Fig. 1.** Simulated spatial patterns of water erosion (A) and tillage erosion (B).

## 3  Application: Comparison of long– and medium–term soil erosion patterns on Loess slopes in Belgium

The potential application of the topographical soil erosion model presented above is illustrated by comparing simulated erosion rates and patterns with estimates of medium and long–term soil erosion for an agricultural field in central Belgium (Van Muysen et al., in press b). The $^{137}Cs$–technique is used to assess medium–term (ie 43 years) erosion (Walling and Quine, 1991) while long–term erosion rates and patterns are deduced from soil profile truncation. The level of truncation is identified by measuring the depth to the upper level of the calcareous loess ($C_2$ horizon). The study site and methods are described in detail in Van Muysen et al. (in press b).

## 3.1   Medium–term soil erosion assessment

It is apparent from Figure 2a that a major part of the field experienced [137]Cs depletion, and consequently, soil erosion during the last 43 years. In order to obtain quantitative soil redistribution rates, the topographical water and tillage erosion model is linked with a [137]Cs–calibration model (Quine 1995, Van Oost et al., submitted). Water and tillage erosion intensities are calibrated until the observed [137]Cs inventories correspond with the simulated inventories (for more details see Van Oost et al., submitted). Application of the calibration procedure resulted in a good fit between simulated and measured [137]Cs inventories ($r^2$=0.53, see Van Muysen et al., in press b).

The redistribution pattern of [137]Cs clearly shows that lower inventories are found on the upslope convex positions and on midslope positions in field A and B. At the same time, a sharp increase in [137]Cs inventory occurs on the transition of footslope to thalweg positions.

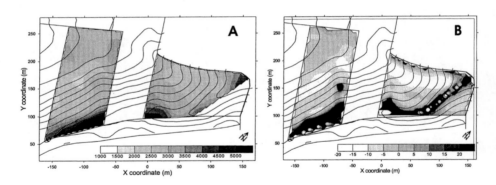

**Fig. 2.** Spatial variation of measured [137]Cs inventories (A) and simulated medium–term (43 years) average soil erosion rates ($t\,ha^{-1}\,yr^{-1}$) (B) (After Van Muysen 2001).

Both water and tillage erosion are contributing significantly to total soil redistribution in the study area. Whereas water erosion is the predominant erosion process on midslope positions, the high erosion rates on the convexities are due to tillage erosion (Figure 2b). The erosion rates on upslope convexities are far more important than erosion rates on midslope positions. Tillage erosion can therefore be seen as the predominant soil erosion process during the last five decades.

## 3.2    Long–term soil erosion assessment

Although some variation may exist, the depth to the calcareous loess allows a quantitative estimate of the spatial distribution of historical erosion and deposition. In Figure 3, the depth to the $C_2$ horizon for field A is presented. Depth to the calcareous loess decreases from the upper plateau positions ($C_2$ material at a depth of 250–300 cm depth) to the steepest slope positions, where $C_2$ material is present directly below and even mixed within the plough layer. Up to more than 300 cm of colluvial deposits are found in the central thalweg positions. This spatial pattern is consistent with the soil truncation that can be expected to have developed on agricultural land when water erosion is considered to be the major cause of soil redistribution: On the plateau position almost no soil erosion has occurred. Erosion is maximized on the steepest and/or longest slopes while deposition is dominant in the relatively flat and wide thalweg. On the other hand, this pattern is inconsistent with the soil redistribution pattern obtained from the $^{137}$Cs–data that indicate that present–day erosion rates are highest on the convex upslope positions where relatively uneroded soil profiles occur. In addition to the deposition in the main thalweg, accumulation of soil material is nowadays also occurring in the small concavity on the slope, where a thin layer of colluvial material is found on top of strongly truncated soils (Figure 3).

## 3.3    Discussion

This indicates that a major change in erosion and sedimentation patterns has occurred over the last decades. The dominance of water erosion over a timescale of several thousands of years explains the spatial pattern of soil truncation. However, the $^{137}$Cs inventories clearly indicate that the present–day pattern of soil erosion and deposition is dominated by tillage. A more quantitative comparison of historical erosion and deposition with recent soil redistribution rates and patterns can be made by comparing the depth to the calcareous loess with the caesium–derived tillage and water erosion rates and patterns. In Figure 4, the depth to the $C_2$ horizon, obtained from the profile description data is plotted against the simulated water ($E_w$) and tillage ($E_t$) erosion and deposition rates, obtained from the caesium–137 data. The resulting pattern corroborates the conclusions of the qualitative analysis: while simulated water erosion rates are highest where the $C_2$ material is found at shallow depths ($< 1$ m), almost no water erosion is simulated for upslope positions where the $C_2$ material is found at depths in excess of c. 2 m. The opposite is true for simulated tillage erosion: the highest erosion rates are calculated for upslope convex positions with relatively uneroded soil profiles, while the most severely truncated soil profiles on linear backslope positions suffer almost no tillage erosion or even some deposition in concave positions. At some footslope locations a high water deposition rate is simulated although the $C_2$ material is found very close to the surface. This is explained by topographical changes: while these positions were loci of strong water erosion in the past they are now experiencing deposition due to the continuous infilling of the thalweg by colluvium and the corresponding decrease of the slope gradient at footslope locations.

The good fit between simulated water erosion rates and observed depth to calcareous loess is also an independent confirmation that the water erosion model produces realistic spatial patterns (see Figure 4): a strong relationship between the depth to the $C_2$ material and water erosion rates is to be expected when water erosion is the predominant soil erosion process.

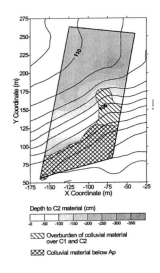

**Fig. 3.** Depth to the calcareous loess material ($C_2$ horizon) (After Van Muysen 2001).

## 4    Representation of space in distributed soil erosion modelling

Various authors have shown that the location of zones of concentrated erosion is determined by the location of runoff generating areas in the landscape, the location of field boundaries and tillage patterns (e.g. Ludwig et al., 1995). Slattery and Burt (1997) and Takken et al. (1999, 2001) demonstrated the effects of field boundaries on sediment deposition and the overall sediment delivery ratio of the watershed. These results emphasize the major effect of landscape structure, i.e. the spatial organization of land units with different land uses and the connectivity between them, on erosion and sedimentation rates and patterns when upscaling from an individual field to a landscape. In this section, a modelling structure (based on WaTEM) is presented that allows assessing the effects of changes in landscape structure on both water and tillage erosion.

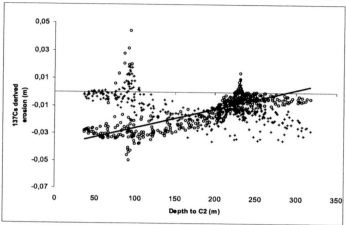

**Fig. 4.** Relationship between $^{137}$Cs derived water (symbol ◯) and tillage (+) erosion rates and the depth of the calcareous loess material (C$_2$ horizon). Regression between $^{137}$Cs derived water erosion and depth to C$_2$: y= 0.0001x-0.0397 (r$^2$=0.69, thalweg locations were excluded in the analysis, see text for explanation). (After Van Muysen 2001).

## 4.1 Model implementation

### In the case of water erosion

On agricultural land, runoff is often directed along field borders and tillage lines instead of the topographic direction (Desmet and Govers, 1997; Takken et al., 2001). Takken et al. (2001) clearly showed that the erosion pattern predicted by a spatially distributed soil erosion model is strongly dependent on the routing procedure used; observed runoff and erosion patterns are simulated much better when field boundaries and tillage direction is accounted for. Secondly, due to the presence of vegetation barriers (hedges, grass strips) and/or because of a change in crop type, overland flow will often be retarded at a field boundary. Furthermore, the infiltration capacity of the vegetation barrier and/or of the lower field can be significantly higher than that of the upper field due to differences in vegetation and/or soil surface condition. Consequently, a fraction of the overland flow will infiltrate near the field boundary and sediment is likely to be deposited here (Meyer et al., 1995; Slattery and Burt, 1997; Takken et al., 1999). The behaviour of the water erosion process at a field boundary is complex and characterized by a high spatial and temporal variability. At present, there is very little information on this topic so that assumptions have to be made. The process may be simplified by assuming that a fixed percentage of sediment and water is trapped at a field boundary representing an average situation over several years.

### In the case of tillage erosion

Unlike water erosion, soil redistribution by tillage will only occur within a field. As such, this may also have important geomorphologic consequences, for each field boundary is a line of zero flux. On sloping land, this leads to important tillage erosion at the downslope side of field boundaries and deposition at the upslope side. Tillage erosion at a field boundary can reach extreme values: even on a gentle 0.01 slope and assuming a $k_{til}$-value of $800\,\mathrm{kg\,m^{-1}}$, the soil loss near a field boundary exceeds $8\,\mathrm{kg\,a^{-1}}$ or about $6\,\mathrm{mm\,a^{-1}}$. In many agricultural areas, the effect of this process can be observed in the field. The incorporation of landscape structure in WaTEM requires knowledge of land use (distinction between arable and non–arable land) and the position of field boundaries. This may be achieved by using a land unit identification file of the study area in which each parcel or land unit has a different identifier. If material is transferred towards a location with a different land unit identifier, specific field boundary conditions are applied according to the erosion process as described above. In Figure 5, predicted spatial patterns of soil erosion on the catchment scale are shown.

### 4.2  Application

Van Oost et al. (2000a) used WaTEM in order to evaluate the effects of changes in landscape structure for three agricultural watersheds situated in central Belgium. Information of the temporal evolution of landscape structure was based on a study of aerial photographs taken in 1947, 1969 and 1990. Figure 5 illustrates the changes in landscape structure for the Kouberg study area. For all study sites, the average field size increased by 200 % to 300 % due to re–allocations of field parcels. An increase in field size normally leads to a higher water erosion risk but this may be more then compensated by the effect of changes in land use. The Ganspoel and Kouberg watersheds experienced an increasing water erosion risk by respectively 12 % and 29 % while Kinderveld experienced a decrease by 15 % between 1947 and 1990 (Figure 6, Table 1). Thus, whether the water erosion risk will increase or decrease depends on the characteristics of the converted land and the degree of the increase in field size.

The evolution of tillage erosion is not only determined by changes in landscape structure but also by the evolution of the tillage transport coefficient. With increased mechanisation and power availability, tillage is carried out deeper and faster, leading to an increase of tillage erosivity (Van Oost et al., 2000a). When changes in land use were relatively small (e.g. Kouberg), tillage erosion rates are found to be similar in 1947 and in 1990. This means that the effect of decreasing tillage erosion at field boundaries, due to the disappearance of field boundaries, was fully compensated by the increase of the tillage transport coefficient. The Ganspoel and Kinderveld watersheds show tillage erosion rates decreasing by respectively 16 % and 33 %. Here, the combined effect of changes in land use and tillage erosion at field boundaries was more important then the increasing tillage transport coefficient.

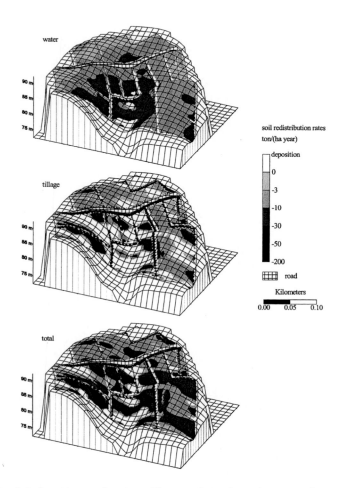

**Fig. 5.** Simulated patterns of water, tillage and total erosion rates for a small area in the Kinderveld watershed (After Van Oost et al., 2000a, published in Landscape Ecology).

**Table 1.** Effect of landscape structure on soil erosion (After Van Oost et al. 2000a).

|  |  | 1947 | 1969 | 1990 | change since 1947 (%) |
|---|---|---|---|---|---|
| Water erosion rate | Ganspoel | 10.2 | 7.9 | 11.4 | **+11.8** |
| $(\mathrm{t\,ha^{-1}\,a^{-1}})$ | Kinderveld | 8.9 | 7.5 | 7.6 | **-14.6** |
|  | Kouberg | 6.6 | 7.9 | 8.5 | **+28.8** |
| Tillage erosion rate | Ganspoel | 11.0 | 8.8 | 9.3 | **-15.5** |
| $(\mathrm{t\,ha^{-1}\,a^{-1}})$ | Kinderveld | 12.4 | 10.9 | 8.3 | **-33.1** |
|  | Kouberg | 8.3 | 9.8 | 8.4 | **+1.2** |
| Total erosion rate | Ganspoel | 19.1 | 14.9 | 18.0 | **-5.8** |
| $(\mathrm{t\,ha^{-1}\,a^{-1}})$ | Kinderveld | 19.6 | 16.7 | 14.2 | **-27.6** |
|  | Kouberg | 13.5 | 15.7 | 14.8 | **+9.6** |

1947                                                  1990

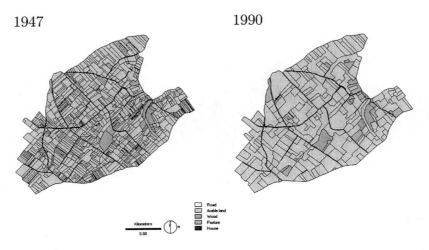

**Fig. 6.** Temporal evolution of field size and land use in the Kouberg catchment for 1947 and 1990.

## 4.3 Discussion

Shifting the focus from homogeneous, individual fields to a landscape as an ecological system, the model results showed the important role of landscape structure. Landscape structure is important to soil erosion because it encompasses the effects of land use and field boundaries. Land use determines the erodibility and hydrological structure of land units while field boundaries regulate the connectivity (transfer of eroded soil and runoff) between different land units. The model predicted changes in mean annual erosion rates up to 28 % due to human induced changes in landscape structure. Nevertheless, the spatial structure of a watershed is strongly simplified in most water erosion models (e.g. WEPP, EUROSEM). Our findings indicate that the way we represent space in distributed models affects model performance at least as dramatically as the physical description of the process.

# 5 Simulating the effects of soil erosion on soil properties

During the last years, various models have been developed to simulate soil erosion rates and patterns. However, loss of fertile topsoil is not the only effect soil erosion has on the soil. Soil erosion will also redistribute soil constituents (e.g. phosphates, organic matter, disease organisms, stones, ?) affecting the spatial distribution of soil properties. Until now, the effect of soil erosion on the redistribution of soil properties cannot be assessed using existing erosion models.

In this section, a mass–balance model linked with WaTEM (Van Oost et al., 2000b ; Van Oost et al., submitted) is used in order to illustrate soil–constituent redistribution by erosion. A full description of the model is beyond the scope of this paper and a detailed description of the model but can be found in Van Oost et al. (2000b) and Van Oost et al. (submitted). In short, a WaTEM–like soil erosion model is used to simulate the horizontal transfers of soil constituents over a landscape. In order to account for the vertical transfers of soil constituents (due to lowering or aggradation of the surface), soil constituent mass–balances are stored for the plough layer and up to four different horizons. The model enables to use an iterative scheme for simulating long–term changes in soil properties. In this case, the resulting spatial distribution at time step t is used as input for time t+1. Two case studies are used to illustrate the potential of the model.

## 5.1 Dispersion of a sand–knoll

The effect of soil erosion on soil properties is illustrated using data collected on the Huldenberg experimental site (Govers, 1987). Here, the sand content (0.063–2.0 mm fraction) was measured for the plough- and sub plough–layer along a slope profile using the wet sieving method. Most of the slope profile is covered by wind–blown loess deposits but a sandy outcrop occurs at the uppermost part (Figure 7). The boundary of this outcrop is well represented in the subsoil by a sharp decrease in sand content between 10 and 12 m. This sharp boundary is absent in the plough–layer: here, the sand content is found to decrease gradually from c. 55 % to c. 10 % over a distance of c. 45 m. The sand content to be expected in a loess–derived topsoil is between 5 and 10 %. Therefore, it could be assumed that the excess sand which is present in the topsoil between 12 and 40 m is not in situ and results from the downslope movement of the sandy material by water and tillage processes after the experimental site was brought into cultivation c. 130 years ago (Govers et al., 1993). The gradual decrease of the sand content over a distance of c. 45 m suggests that important dispersion took place while the sandy material was transported over the silty loamy subsoil.

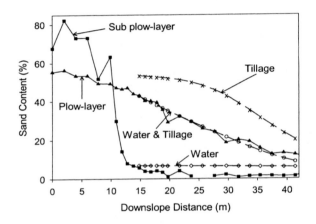

**Fig. 7.** Measured sand content of plough-layer (▲) and sub plough–layer (■) on the Huldenberg field. The measured sand content is compared with three model predictions where only water (◊), only tillage (×) and water and tillage (○) processes are taken into account (after Van Oost et al., 2000b).

In order to investigate the relative role of tillage and water in the genesis of the observed pattern, the model, as described above, was used.

Soil redistribution by diffusive processes like soil creep and rain splash is ignored since their intensity is much lower than water and tillage erosion (Govers et al., 1994). The model is implemented with the following assumptions (i) the sandy outcrop was already present when the field was tilled for the first time and the sand content of the plough layer in this area has remained constant (55 %), (ii) the silty loam has an initial sand content of 10 % for the plough–layer and 3 % for the sub plough–layer, (iii) tillage depth was constant and equaled

0.15 m and (iv) for the water erosion component of the model, the parameter values proposed by Govers et al. (1993) were used.

Tillage conditions over the last 130 years are unknown: tillage intensity is therefore calibrated. Variations in the parameter values of the tillage erosion model did not significantly affect the predicted spatial pattern.

Table 2 and Figure 7 show that the agreement between observed sand content and model output is poor when only water processes are considered (simulation 1). Overland flow reduces the sand content of the plough layer in the silty loam area which is due to the fact that eroded soil particles are transported downslope until the transport capacity is exceeded: as in our case the slope is convex, all material eroded by water is simulated to leave the slope at the bottom end. Due to the lowering of the surface by erosion, subsoil poor in sand is continuously mixed into the plough layer. When only tillage is taken into account (simulation 2), the correspondence between predicted and observed values of sand content is better and a clear downslope movement of sand–rich topsoil over the silty subsoil is simulated. However, the simulated sand content is always above the observed values. Apparently, dispersion due to tillage alone does not lead to the very gradual decrease of sand content observed in the field. In simulation 3, tillage and water are combined, resulting in a much better correlation between observed and predicted redistribution patterns. The transition between sandy and silty topsoil is now much more gradual due to combination of two processes: (i) soil tillage continuously moves sand downslope and (ii) at the same time sand dilution of the sand occurs due to the mixing of subsoil poor in sand with the plough layer as a result of surface lowering by both water and tillage erosion. Optimal linear regression coefficients for the tillage erosion model are given in Table 2 and are in close agreement with the experimentally derived data in combination with four tillage operations per year. These values correspond to a $k_{til}$–value of $324 \, \text{kg m}^{-1} \, \text{yr}^{-1}$ which is higher, but in the same order of magnitude, then the $k_{til}$ –value of $133 \, \text{kg m}^{-1}$ as found by Govers et al. (1994) based on height differences. It is recognized that the procedure used here is not a rigorous model validation. However, this exercise merely seeks to point out the typical characteristics of tillage and water processes and it is felt that the assumptions are adequate for this purpose.

## 5.2   Modelling historical sediment production and topsoil properties

When modelling soil erosion in a historical perspective, it is important to account for the spatial and temporal variations in soil properties. Nachtergaele and Poesen (in press) have clearly shown that the resistance of different soil horizons vary significantly. Based on flume experiments, they concluded that the Ap and C horizons of a loess–soil are respectively 1.7 and 16 times more erodible then a Bt under average soil moisture conditions. Since soil profiles are truncated due to water and tillage erosion, spatial variation in soil erodibility can therefore be expected. Consequently, when modelling long–term soil erosion rates, the use of a single erodibility factor will lead to serious errors on predicted erosion volumes and patterns (Nachtergaele and Poesen, in press).

**Table 2.** Comparison of measured and predicted sand content using only water, only tillage and a combination of both processes. (After Van Oost et al., 2000b).

| | | | | Model validation | | |
| Simulation No. | Water | Tillage | SSQ[a] | Gradient[b] | Intercept | $R^2$ |
|---|---|---|---|---|---|---|
| 1 | Yes | No | 496 | 34.7 | -200.1 | 0.88 |
| 2 | No | Yes | 286 | 0.84 | -9.7 | 0.82 |
| 3 | Yes | Yes | 4 | 0.90 | 2.9 | 0.97 |

[a]SSQ is the sum of squares of differences between measured and predicted sand content divided by the number of observations (n=17). [b]Regression of measured on predicted sand content.

WaTEM was applied to the field described in section 3 and soil erosion was modelled since the start of cultivation, approximately 1500 to 2000 years ago. It was assumed that initially, the field had a typical soil profile, which can be found at present under undisturbed circumstances (Figure 8). With continuing truncation of the soil after the start of cultivation, soil properties will change: (i) incised water erosion may reach other horizons and (ii) the underlying soil horizons will be mixed into the plough layer and consequently, the properties of the topsoil will change.

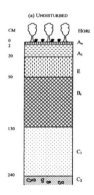

**Fig. 8.** Typical soil profile, developed in soil under deciduous forest (After Van Muysen 2001).

In Figure 9 the results of the model runs are presented. Figure 9a shows the patterns and rates of soil truncation due to water erosion after 1000 years. On the plateau, little soil truncation occurred so that topsoil properties remained relatively constant. On the other hand, on the slope positions the B–horizon is mixed into to plough layer resulting in a high clay content (Figure 9b). Most erosion is simulated in the thalweg where overland flow concentrates. Here, erosion has lead to the complete removal of the $A_p - B_t$ complex so that the characteristics of the C–horizon are reflected in the plough–layer.

From the result of a soil profile study (Van Muysen et al., in press b), it was concluded that water erosion was the dominant process until the mechanization of agriculture. However, since the start of cultivation, agricultural fields have been cultivated for long time with an ard–type of plough. Tillage experiments conducted with this type of plough in developing countries show that the transport coefficient can range up to $250\,\mathrm{kg\,m^{-1}\,yr^{-1}}$ which is approximately one third of the present day tillage intensity ($\pm\,700\,\mathrm{kg\,m^{-1}\,yr^{-1}}$) (e.g. Nyssen et al., 2000). It may be questioned whether the tillage systems, used before the intensification of agriculture, may be ignored in long–term erosion studies. A new simulation was therefore carried out: the effect of tillage was implemented assuming an average tillage intensity of only $140\,\mathrm{kg\,m^{-1}\,yr^{-1}}$. In Figure 9c, soil truncation due to the combined effect of water and tillage erosion is presented. The pattern is still dominated by water erosion but is different from the pattern that is predicted from water erosion alone: In the thalweg, tillage deposition leads to a reduced soil truncation in the upper part of the thalweg. On the other hand, tillage erosion on the convexities leads to increased truncation. This is also reflected in the clay content of the topsoil (Figure 9d).

This clearly has implications for ephemeral gully erosion as well: a gully may reach the highly erodible C–horizon much earlier in the case where only water erosion is active. When tillage is applied, tillage deposits will protect the C–horizon in the thalweg.


# 6 Discussion and conclusion

In this paper, the possibilities of topography–based, spatially distributed soil erosion models are illustrated. These models try to capture the essence of reality while only using topographical parameters to represent a variety of soil redistribution processes. When simulating soil erosion processes over longer periods of time, we believe that this approach is more appropriate then sophisticated, physically based erosion models (e.g. WEPP, EUROSEM). These complex models require a large amount of input data. For example, the simplest model requirements of the WEPP slope–profile model demand as a minimum some 100 input parameters. This will lead to major problems with respect to spatial and temporal variability of parameter values and uncertainty of parameter estimates and hence results. In contrast, the topographical models presented in this paper, have a limited number of parameters which will minimize optimization problems and is likely to make the final optimized values more physically meaningful (e.g. Quinn and Beven, 1993). Comparison of model simulations with erosion rates derived from soil profile truncation showed the potential of this approach. However, the model presented here is a steady state model and the basic assumption is therefore uniformitarianism. A dynamic approach may be required to assess the effects of extreme events.

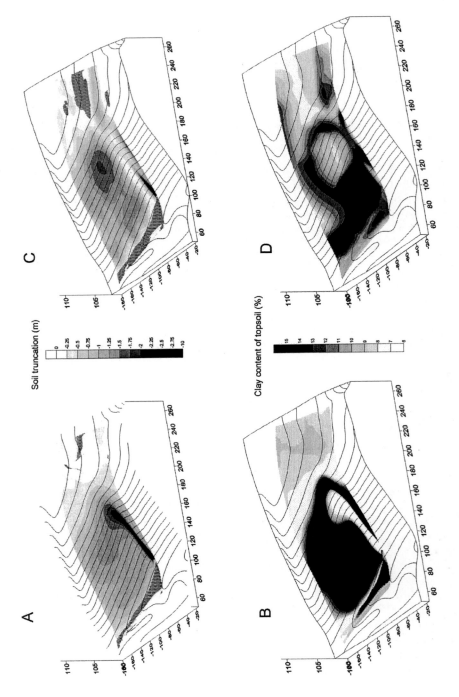

**Fig. 9.** Simulated soil truncation after 1000 years applying a water erosion scenario (A & B) and a combined water and tillage erosion scenario (C& D).

In this paper, attention was also given towards representation of the landscape in spatially distributed soil erosion models. Field observations clearly showed that field borders may act as water and sediments collectors and water flow may be routed along linear features, such as lynchets or roads. Therefore, total sediment yields and the actual pattern of runoff may be very different from the yields and patterns that would be predicted from topography alone. Applying WaTEM to a catchment where significant changes in landscape structure occurred during the last decades, indicated that when estimating environmental change impact on soil erosion at larger scales, modelling approaches that explicitly deal with the spatial aspects of soil erosion must be used. It was concluded that the way space is represented in distributed models affects model performance at least as dramatically as the physical description of the process.

Comparison of historical soil erosion rates with medium–term soil erosion rates indicated that an important shift in the relative contribution of water and tillage erosion rates to total soil redistribution on agricultural land has occurred in the last decades. While water erosion has been dominant since the start of cultivation, tillage erosion has become an at least as important soil redistribution process during the last decades. This is due to the increased mechanisation in agriculture. However, this does not mean that tillage erosion processes are not relevant for long–term soil erosion studies. However, although tillage intensity has been relatively low until a few centuries ago, it is a process affecting relief morphology and the spatial pattern of soil properties significantly. Soil tillage is therefore an important key when studying soil profiles and sedimentary records. The most visible effect of tillage is the formation of soil banks at parcel borders, affecting relief morphology and landscape structure.

Finally, most soil erosion models focus on the prediction of sediment production and sediment export to rivers. Until now, very little attention is given towards the spatial and temporal variation in soil properties due to soil erosion. On the long–term, soil profiles are truncated due to water and tillage erosion and spatial variation in outcropping soil horizons is observed. Model applications whereby a topographical soil erosion model is linked with a mass–balance model showed that soil erosion is a highly spatial and temporal variable process. It was concluded that implementing this variability represents a vital element when trying to link surface processes with sedimentary records.

# References

De Alba, S. (in press): Modelling the effects of complex topography and patterns of tillage on soil translocation by tillage with mouldboard plough. A three–dimensional approach. Soil and Tillage Research.

Desmet, P.J.J., and Govers, G. (1996): A GIS procedure for automatically calculating the USLE LS factor on topographically complex landscape units. J. of Soil and Water Conservation 51: 427–433.

Desmet, P.J.J., and Govers, G. (1997): Two–dimensional modelling of the within–field variation in rill and gully geometry and location related to topography. Catena 29: 283–306.

Govers, G. (1987): Spatial and temporal variation in rill development processes at the Huldenberg experimental site. In Bryan, R. (eds): Rill erosion: Processes and significance. Catena Supplement 8: 17–34.

Govers, G., and Poesen, J. (1988): Assessment of the interrill and rill contributions to total soil loss from an upland field plot. Geomorphology 1: 343–354.

Govers, G., Vandaele, K., Desmet, P.J.J., Poesen, J., and Bunte, K. (1994): The role of soil tillage in soil redistribution on hillslopes. European Journal of Soil Science 45: 469–478.

Govers G., Quine, T.A., and Walling, D.E. (1993): The effect of water erosion and tillage movement on hillslope profile development: a comparison of field observations and model results. In Farm Land Erosion in Temperate Plains Environments and Hills. 587 pp. Edited by S. Wicherek, Elsevier, Amsterdam.

Ludwig, B., Boiffin, J., Chadoeuf, J., and Auzet, A.V. (1995): Hydrological structure and erosion damage caused by concentrated flow in cultivated catchments. Catena 25: 227–252.

McCool, D.K., Foster, G.R., Mutchler, C.K., and Meyer, L.D. (1989): Revised Slope Length Factor for the Universal Soil Loss Equation. Transactions of the ASAE 32(5), 1571–1576.

Meyer, L.D., Dabney, S.M., and Harmon, W.C. (1995): Sediment–Trapping Effectiveness of Stiff–grass Hedges. Transactions of the ASAE 38(3): 809–815.

Nachtergaele, J., and Poesen, J. (in press): Spatial and temporal variation in resistance of loess–derived soils to ephemeral gully erosion. European Journal of Soil Science.

Nyssen, J., Poesen, J., Mitiku, H., Moeyersons, J., and Deckers, J., (2000): Tillage erosion on slope with soil conservation structures in the Ethiopian highlands. Soil and Tillage Research 57: 155–127.

Quine, T.A. (1995): Estimation of erosion rates from Caesium–137 data: the calibration question. In: Foster, I.D.L., Gurnell, A.M., and Webb, B.W. (eds.): Sediment and Water Quality in River Catchments. John Wiley and Sons, Chichester: 307–329.

Quinn P.F., and Beven K.J. (1993): Spatial And Temporal Predictions Of Soil–Moisture Dynamics, Runoff, Variable Source Areas And Evapotranspiration For Plynlimon, Mid–Wales. Hydrol Process 7: (4) 425–448.

Slattery, M.C., and Burt, T.P. (1997): Particle size characteristics of suspended sediment in hillslope runoff and stream flow. Earth Surface Processes and Landforms 22: 705–719.

Takken, I., Beuselinck, L., Nachtergaele, J., Govers, G., Poesen, J., and Degraer, G. (1999): Spatial evaluation of a physically–based distributed erosion model (LISEM). Catena 37: 431–447.

Takken, I., Govers, G., Jetten, V., Nachtergaele, J., Steegen, A., and Poesen, J., (2001): Effects of tillage on runoff and erosion patterns. Soil and Tillage Research 61: 55–60.

Van Muysen, W., Govers G., Bergkamp, G., Poesen J., and Roxo, M. (1999): Measurement and modelling of the effects of initial soil conditions and slope gradient on soil translocation by tillage. Soil and Tillage Research 51: 303–316.

Van Muysen, W. (2001): Tillage translocation and tillage erosion: an experimental approach. Unpublished PhD Thesis. Katholieke Universiteit Leuven.

Van Muysen, W., Govers, G., and Van Oost, K. (in press a): Identification of important factors in the process of tillage erosion: the case of mouldboard tillage. Journal of Soil and Water Conservation.

Van Muysen, W., Govers, G., Van Oost, K., Deckers, J., and Raabe, R. (in press b): Comparison of long and medium–term soil erosion rates and patterns on loess slopes in central Belgium: from water erosion dominated to tillage erosion dominated landscape evolution. Catena.

Van Oost K., Govers, G., and Desmet, P. (2000a): Evaluating the Effects of Changes in Landscape Structure on Soil Erosion by Water and Tillage. Landscape Ecology 15: 577–589.

Van Oost K., Govers, G., Van Muysen, W., and Quine, T.A. (2000b): Modeling Translocation and Dispersion of Soil Constituents by Tillage on Sloping Land. Soil Science Society of America Journal 64: 1733–1739.

Van Oost, K., Van Muysen, W., Govers, G., Heckrath, G., and Quine, T.A. (in press): Simulation of Soil Constituent Redistribution by Tillage on 2D Complex Topographies. European Journal of Soil Science.

Walling, D. and Quine, T.A. (1991): Use Of $^{137}$Cs Measurements To Investigate Soil–Erosion On Arable Fields In The UK – Potential Applications And Limitations. Journal of Soil Science 42: 147–165.

# Long–term and large scale high resolution catchment modelling: Innovations and challenges arising from the NERC Land Ocean Interaction Study (LOIS)

Tom J. Coulthard and Mark G. Macklin

Institute of Geography and Earth Sciences, University of Wales, Aberystwyth, Wales, SY23 3DB, UK

**Abstract.** The CAESAR (Cellular Automaton Evolutionary Slope And River) model is used to simulate the Holocene evolution of four major tributaries of the Yorkshire Ouse (The Rivers Swale, Ure, Nidd and Wharfe). Using a regular grid to represent each river catchment, the model simulates erosion and deposition for every flood over the last 9000 years driven by a land cover history derived from palynological sources and a rainfall record reconstructed from peat bog wetness indices. Results from these simulations show that all four catchments demonstrate rapid increases of sediment discharge in response to wet shifts in the climate record, and the magnitude of sediment yields are amplified after simulated catchment deforestation. However, there are some differences in sediment yield between the river catchments which are caused by sediment storage and re–mobilisation within lower gradient reaches or sub basins. This modelling study shows how it is possible to simulate the evolution of large river catchments over Holocene time scales, but highlights that there are still many problems to be overcome, especially with process representation.

## 1 Introduction

This paper draws builds upon research carried out under phase 2 of the Land–Ocean Evolution Perspective Study (LOEPS), a major component of the Land Ocean Interaction Study (LOIS) which ran from 1991 to 1998. The LOIS program and associated sub projects was an inter–disciplinary, multi million pound research project funded by the UK Natural Environment Research Council (NERC) (Shennan and Andrews, 2000). Its principal aims were to establish contemporary fluxes of sediments, nutrients and contaminants to and from the coastal zone; to characterise the key physical geochemical and biological processes operating there; to describe the Holocene evolution of coastal and river systems; and to develop coupled land–ocean models to simulated the transport and fate of materials in the coastal zone with the ability to predict future changes. It was based on the eastern and north eastern coast of England, from Berwick upon Tweed to Great Yarmouth (Shennan and Andrews, 2000).

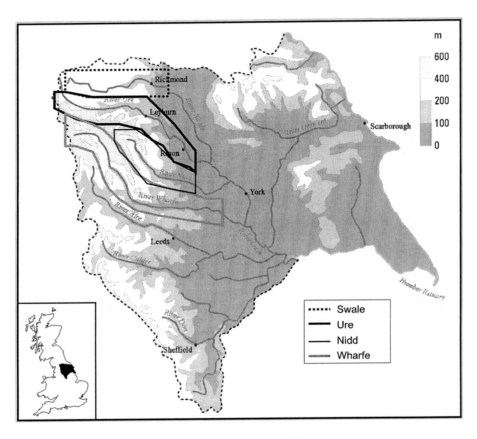

**Fig. 1.** Location of study catchments.

A central element of LOEPS was to determine Holocene river sediment fluxes for the Yorkshire Ouse catchment (Figure 1). This was carried out by using geomorphological surveys calibrated with $^{14}C$ dates (Macklin et al. 2000) and through the development of a new numerical model called CAESAR (Cellular Automaton Evolutionary Slope And River model; Coulthard and Macklin, 2001; Coulthard et al., 2002). Designed to simulate river basin response to environmental change, this model is catchment based, fully integrating slope and fluvial processes at a high spatial resolution (50 m) and temporal scale (simulating individual flood events). During the LOIS project, CAESAR was used to simulate the evolution of the River Swale, a major tributary of the Yorkshire Ouse (Figure 2), over the last 9000 years (Coulthard and Macklin, 2001). Driven by proxy records of land use and climate, this simulation showed how the catchment responded rapidly to wetter climates, with significant increases in sediment delivery. Peaks in sediment discharge closely followed wet shifts in climate, however, land cover changes and sediment storage also affected the size of these peaks. Importantly, these results were validated by comparison to an independent data source – the frequency of dated British Holocene alluvial units (Figure 2). However, this paper only examined how one tributary of the Yorkshire Ouse (the River Swale) responded to climate and land cover changes. This leaves an important question, how do different but contiguous catchments respond to environmental change? Furthermore, if there is a different response from these catchments, is it caused by the catchment morphology, internal thresholds, sediment storage and remobilisation, or by the model itself?

To answer these questions, three major tributaries of the Yorkshire Ouse, the Nidd, Ure and Wharfe, were modelled and the results compared to each other and the Swale simulations. This paper briefly describes the model, how all four catchments were simulated; the results and implications from these model runs, and finally what experience from this study could be applied to the RhineLU-CIFS project (Lang et al., 2000 ).

## 2   The CAESAR model

The CAESAR model (full details can be found in Coulthard et al., 2002) represents a river catchment with a mesh of uniformly sized square grid cells. Each grid cell contains values for elevation, water discharge, vegetation cover and grain size distribution. To model convergent and divergent flow, a scanning algorithm is used that sweeps across grid cells four times (from top, bottom, left and right). In each scan, flow from a hydrological model is routed to any (or all) of the three downstream immediate neighbours as per Murray and Paola (1994). However, unlike Murray and Paola's method, if flow exceeds the cells subsurface hydraulic conductivity, water depth is calculated with a modification of Manning's formula (Coulthard et al., 2000) using the average positive slope from all eight neighbours. If the combined elevation of the cell, and its water depth, is greater than one (or all) of the three downstream neighbours then a proportion is routed to these cells. The maximum depth calculated for all cells

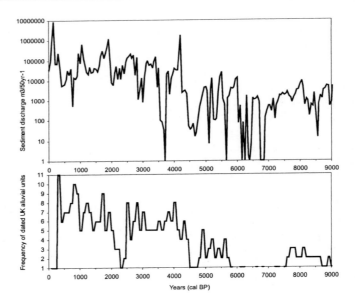

**Fig. 2.** Simulated sediment yield for the river Swale, compared to frequency of dated British alluvial units.

over each of the four scans is recorded and has been shown to be similar to calculations using multiple flow algorithms (Desmet and Govers, 1996). Any flow not removed from the basin is left, allowing hollows to fill up so they do not to interfere with subsequent scans. Whilst unable to represent momentum effects or secondary circulation, this procedure is several orders of magnitude faster than other methods and most importantly allows convergent and divergent flow in all directions, over local topographic highs, while maintaining mass balance and fluxes along the maximum energy gradient.

Flow depth and bed slope are then used to calculate fluvial erosion between cells through the Einstein–Brown (1950) expression. Sediment transport is calculated for eleven separate grain size fractions (in whole phi units from 0.001 to 0.256 m) coupled to the stream bed through eleven active layers (Hoey and Ferguson, 1994; Cui et al., 1996) each 0.2 m thick, creating a 2.2 m stratigraphy for each cell. Mass movement occurs when a slope threshold is exceed, soil creep is also incorporated and a simple linear vegetation growth model allows a turf mat to develop. There is no explicit distinction between whether a cell is a 'slope' or 'channel' cell, simply if there is enough water to generate overland flow, then fluvial erosion and deposition is possible. Changes in elevation resulting from fluvial erosion, mass movement and soil creep are updated simultaneously, and a variable time step is used to restrict net erosion to 10 % of the local slope preventing computational instability. As all these processes are operating within the same regular grid, feedbacks between them are automatically integrated, allowing inputs of rainfall, vegetation cover and topography to drive landscape evolution.

# 3   Study catchments and initial conditions

The Swale, Ure, Nidd and Wharfe are all major tributaries of the Yorkshire Ouse draining in an easterly direction from the Yorkshire Dales in to the Vale of York (Figure 1). The geology of the catchments is dominated by coarse sandstones and gritstones of the Yoredale series and Carboniferous Limestone. Relief and drainage area of the modelled sections are listed in Table 1.

**Table 1.** Table of tributary catchment characteristics.

| River | Swale | Ure | Nidd | Wharfe |
|-------|-------|-----|------|--------|
| Modelled drainage area (km2) | 383 | 646 | 281 | 697 |
| Max. relief of modelled area (m) | 514 | 564 | 560 | 546 |
| Main channel length (km) | 62 | 82 | 49 | 97 |

For each catchment simulation of an initial topography, climate record and land cover history were required. As the precise valley topography of the four catchments at the beginning of the Holocene is unknown, the present day surface was taken as an analogue. This assumption is not unreasonable as following large scale valley floor incision (10–20 m) during the Late Glacial (Howard et al., 2000) trunk channel bed levels have varied by only $\pm 3$ m over the Holocene (Macklin et al., 2000). This initial topography of each catchment was represented by a 50 by 50 m digital elevation model (DEM). This DEM was overlain with a 3 m thick 'soil' of homogenous grain size distribution capped with a turf mat. The climate input for the model is derived from a combination of two proxy wetness indices from peat bogs in northern England (Bolton Fell Moss; Barber et al., 1994) and northern Scotland (Anderson et al., 1998). We have used the nearer Bolton Fell Moss record (located 60 km north west of the River Swale) back to 6300 cal. BP and the Scottish record, which starts at 9200 cal. BP, for the early Holocene. This combined sequence was re–sampled at 50 year intervals, and based on previous applications (Coulthard et al. 2000), normalised to values between 0.75 and 2.25 (Figure 3) to create a rainfall index. To drive the model, a 10 year hourly rainfall record (1985–1995) located in the lower part of the Swale catchment was duplicated 5 times to span 50 years, and multiplied by the index, generating a proxy hourly rainfall record for the last 9000 years. Changes in land cover are poorly documented in the Swale catchment and we have used local palynological records (Tinsley, 1974; Smith, 1986) to develop a land cover index ranging from 2 (forested) to 0.5 (grassland); (Figure 3). This simulated the effects of different land–cover on catchment hydrology by altering a parameter, within the hydrological model, controlling the magnitude and duration of the flood hydrograph for a given storm event (Coulthard et al. 2000). Identical climate and land cover data sets were applied to all four catchments. During the simulation, for each catchment continuous sediment discharge data for the grain size fractions were recorded and grid elevation and grain size data were saved at 50 year intervals. From these sediment discharge records were produced (Figure 3) and cumulative percentages of sediment yield were calculated (effectively normalising the results across all four catchments).

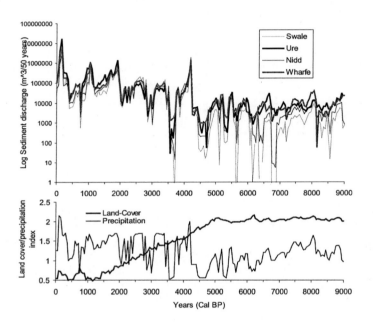

**Fig. 3.** Simulated sediment yield for the Rivers Swale, Ure, Wharfe and Nidd, plotted with the land cover and precipitation index used to drive the model.

## 4   Holocene river sediment yield simulations

Modelled river sediment discharges (Figure 3) clearly show that for all four catchments the Holocene is characterised by short (10–100 years) periods of high sediment discharges. These peaks correspond closely to a wetter climate indicating that climate is the main driver of changes in sediment yield for all of these catchments. However, land cover changes, principally in the form of tree clearance, appear to have an important effect, as in all four catchments the peaks in sediment discharge after 2000 cal. BP are significantly larger in magnitude than those prior to deforestation. This is also clearly illustrated in the cumulative percentage plot of sediment yield (Figure 4) that shows all four catchments have a far smaller increase in sediment yield in response to the wetter periods under forested cover (c. 3100–3400 cal. BP and c. 2500–2750 cal. BP) than those under deforested conditions (c. 1800 cal. BP and c. 1000 cal. BP). For example, closer examination of Figure 4 shows that the River Ure records only a 3 % increase in sediment yield for the period 4000 to 1900 cal. BP when forested but a 25 % increase from 1900 to 400 cal. BP following tree clearance. This trend, which is replicated in the other three catchments, implies that reduced tree cover influences catchment response to climate changes by increasing the amplitude of the sediment peak for a given storm event.

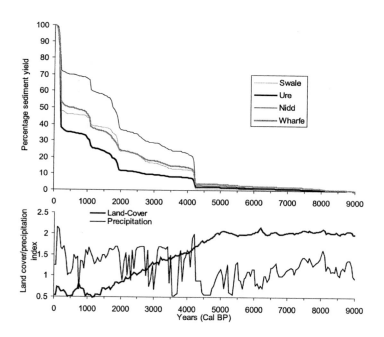

**Fig. 4.** Cumulative simulated sediment yield for the Rivers Swale, Ure, Wharfe and Nidd, plotted with the land cover and precipitation index used to drive the model.

Figure 3 also shows, however, that there are significant differences between how the four tributaries respond to identical climate and land cover changes. The smaller Rivers Swale and Nidd (Table 1) both have a far more 'flashy' spiky sediment yields during the Holocene compared to the larger Ure and Wharfe that have less variable sediment yields. The cumulative percentages (Figure 4) also reveal very different behaviour from the catchments. The cumulative sediment discharge from the Nidd rises by 50 % between c. 4200 cal. BP and 400 cal.

BP, but during the same period the larger Ure increases by only 25 %. Interestingly, the Swale and Wharfe have very similar sediment discharge characteristics, despite having a considerable difference between catchment areas. This would indicate that whilst total sediment yield may reflect the catchment area, the nature and the timing of river sediment discharge is not solely controlled by size.

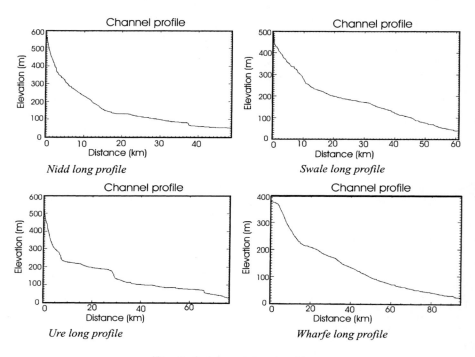

**Fig. 5.** Catchment long profiles.

Analysis of channel long profiles (Figure 5) indicates that differences in the magnitude and to some degree the timing of sediment discharge peaks may be primarily controlled by catchment morphology. The Nidd falls rapidly from 560 m to 52 m over 49 km and this may drive the consistently high sediment yield. Conversely, the Ure drops rapidly in its headwaters down to 18 km and has a number of major steps in its long profile that separate lower gradient alluvial basins. These are long–term sediment stores and material is only re–mobilised and exported downstream by major climate related changes in flood frequency and magnitude, such as during the Little Ice Age (LIA). Despite having different channel lengths, the Swale and Wharfe have similarly shaped long profiles and this may in part result in the comparable sediment yields (Figure 5). Figure 6 shows the hypsometric curves for the four catchments and there seems to be no relationship between the curves and the cumulative sediment yields. This implies that it is the gradient and characteristics of the channel network rather than the distribution of uplands and lowlands that causes changes between the four simulated catchments.

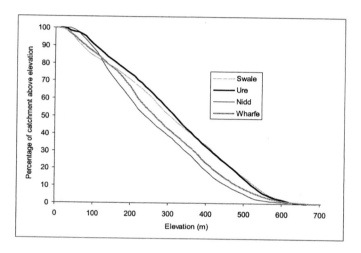

**Fig. 6.** Hyposmetric curves for the modelled sections of the Rivers Swale, Ure, Nidd and Wharfe.

Previous work on the River Swale (Coulthard and Macklin, 2001) had suggested that different sediment yields produced by similar sized climate changes were the result of sediment supply and storage effects. This new analysis would appear to again highlight the importance of basin storage effects and catchment morphology in controlling sediment yields.

## 5   Innovations and challenges for river basin modelling

The CAESAR based modelling work carried out under the LOIS project provided several innovations. First and foremost, whole catchments were modelled at a high spatial resolution, simulating every flood during the 9000 year model runs. It integrated slope, channel and hydrological processes within one framework and has been validated by independent data. This was the first time in the UK that real catchments have been simulated at this resolution over Holocene time scales, and has provided new insights into how river systems respond to environmental change.

These studies have also revealed several challenges for river basin modelling. Despite the relatively high resolution modelling of basins several hundred km$^2$ in drainage area, the catchments are all represented by 50 m grid cells. In many places the river channels are smaller than 50 m and the topography more complex than can be accurately depicted with these grid cells. CAESAR has been applied to smaller catchments with 2–3 m grid cells (Coulthard et al., 2000; Coulthard et al., 2002) but there is a trade off between resolution and the size of catchment that can be studied. However, the rapid increase in computational power means that in the future, CAESAR will be able to model similar size catchments at a finer resolution or study even larger basins. Another problem highlighted by these simulations was collecting suitable data to drive them. For the climate record we were fortunate enough to have access to a wetness index derived from raised mires (Barber et al., 1994; Anderson et al., 1998) which was in turn used to modify a rainfall record. But this only changes rainfall magnitude and not frequency, which is clearly not a true picture of how changes in wetness occur. A more realistic solution would be to use a stochastic generation of storm events which could be altered by the wetness record. Land cover changes were far harder to quantify and as a consequence, we were forced to reconstruct an approximate history of deforestation in these catchments based upon palynological and archaeological data. Vegetation changes were applied over the whole catchment, which is clearly an over simplification and runs are required to examine how deforesting certain parts of the catchment will effect sediment delivery.

Many of the challenges arising from the LOIS project relate to scale and process representation. The fundamental problems of modelling large catchments, which are of direct relevance to the RhineLUCIFS study, is to integrate as many relevant processes operating at a fine as possible scale at a high temporal resolution. There are several ways of approaching these issues. Looking at the spatial scale, it is possible to use an irregular sized mesh of nodes and links (e.g. Braun and Sambridge, 1997; Tucker et al., 2001) where topographically simple areas of the catchment are represented with relatively few nodes and links (e.g. a simple hill slope) whereas complex areas (e.g. a river channel) will contain a large number. An alternative is to average the effect of a small scale process across a large grid cell, for example by averaging soil erosion rates over 100 m grid cells. However this coarse grid cell size may conceal rills and small gully's within the cell, which could significantly enhance soil erosion in that area. Another technique may be to use a larger number of regular cells, as used by the CAESAR model, or to develop parallel computing techniques so that several sub–catchments can be modelled in parallel.

With regard to processes being modelled, it is an obvious truism that relevant ones should be included. However, deciding what processes are relevant is not a simple task, especially when modelling landscape evolution over large areas. For example, rock fall may be important in mountainous headwaters, yet irrelevant in lowland areas. To further complicate matters, there are processes or influences about which we know very little, yet may be of vital importance to the models outcome. One example of this is how to include anthropogenic changes on the environment. We have a fairly good documented record of recent changes in agriculture and deforestation over the last 100 to 200 years, and how these have effected river catchments. But we cannot be certain of the timing and location of previous changes – which will have a significant impact. Further-more, if we wish to simulate future changes how are these integrated? Statistical representations of future climates can be generated, but how can we account for altering agricultural practices that may be triggered by legislation or EU subsidy changes?

Additionally, the way in which these processes are represented and to what level of detail are important. For example, soil creep may be described by long–term averages or movement rates, but rill development may require a more detailed event based representation. This will also depend upon the temporal scale of study as soil creep may be negligible when simulating short periods over 10 years, conversely precise details of rill location may become irrelevant when simulating river evolution over the Pleistocene. Possible solutions to these problems include nesting sub models with the framework of a larger model, or by finding new methods to represent them such as fractal distributions or neural networks.

## 6   Conclusions

CAESAR model simulations have shown that the long–term effects of environmental change on large river catchments can be modelled, providing us with useful new insights into how these systems may respond to climate and land–cover changes. There are many challenges involved in modelling large river catchments and the success of future projects may stand or fall on what processes are modelled and how.

### Acknowledgments

This research was supported by Natural Environment Research Council (NERC) grant GST/02/0758.

## References

Anderson, D. E., Binney, H. A., and Smith, M. A. (1998): Evidence for abrupt climatic change in northern Scotland between 3900 and 3500 calendar years BP. The Holocene, 8:97–103.

Barber, K. E., Chambers. F. M., Maddy. D., Stoneman. R., and Brew. J. S. (1994): A sensitive high resolution record of late Holocene climatic change from a raised bog in northern England. The Holocene, 4(2): 198–205.

Braun, J., and Sambridge, M. (1997): Modelling landscape evolution on geological time scales: a new method based on irregular spatial discretization. Basin Research, 9:27–52.

Coulthard, T. J., Kirkby, M. J., and Macklin, M. G. (2000): Modelling geomorphic response to environmental change in an upland catchment. Hydrological Processes, 14:2031–2045.

Coulthard, T. J., and Macklin, M. G. (2001): How sensitive are river systems to climate and land–use changes? A model based evaluation. Journal of Quaternary Science, 16:346–351.

Coulthard, T. J., Macklin, M. G., and Kirkby, M. J. (2002): A cellular model of Holocene upland river basin and alluvial fan evolution. Earth Surface Processes and Landforms, 27(3): 269-288.

Cui, Y., Parker, G., and Paola, C. (1996): Numerical simulation of aggradation and downstream fining. Journal of Hydraulic Research, 34(2): 185-204.

Desmet, P. J. J., and Govers, G. (1996): Comparison of routing algorithms for digital elevation models and their implications for predicting ephemeral gullies. International Journal of Geographical Information systems, 10(3): 311-331.

Einstein, H. A. (1950): The bed–load function for sediment transport on open channel flows. Tech. Bull. No. 1026, USDA, Soil Conservation Service, 71.

Hoey, T., and Ferguson, R. (1994): Numerical simulation of downstream fining by selective transport in gravel bed rivers: Model development and illustration . Water resources research, 30(7): 2251-2260.

Howard, A. J., Macklin, M. G., Black, S., and Hudson–Edwards, K. A. (2000): Holocene river development and environmental change in Upper Wharfedale, Yorkshire Dales, England. Journal of Quaternary Science, 15: 239-252.

Lang, A., Preston, N., Dikau, R., Bork, H.-R., and Mäckel, R. (2000): Land Use and Climate Impacts on Fluvial Systems During the Period of Agriculture – Examples from the Rhine catchment. PAGES Newsletter, 8/3: 11–13.

Macklin, M. G., Taylor, M. P., Hudson–Edwards, K. A., and Howard, A. J. (2000): Holocene environmental change in the Yorkshire Ouse basin and its influence on river dynamics and sediment fluxes to the coastal zone.–In: Shennan, I., and Andrews, J. E. (eds): Holocene land–ocean interaction and environmental change around the western North Sea. Geological Society, Special Publications, 166: 87–96.

Murray, A. B., and Paola, C. (1994): A cellular model of braided rivers. Nature, 371: 54–57.

Shennan, I., and Andrews, J. (2000): An introduction to Holocene land–ocean interaction and environmental change around the western North Sea.–In: Shennan, I., and Andrews, J. (eds): Holocene land–ccean interaction and environmental change around the western North Sea. Geological Society, London. Special Publications, 166: 1–7.

Smith, R. T. (1986): Aspects of the soil and vegetation history of the Craven District of Yorkshire.–In: Manby, T. G., and Turnbull, P. (eds): Archaeology in the Pennines. B.A.R. British Series 158, Oxford: 3–28.

Tinsley, H. M. (1975): The former woodland of the Nidderdale Moors (Yorkshire) and the role of early man in its decline. Journal of Ecology, 6: 1–26.

Tucker, G.E., Lancaster, S.T., Gasparini, N.M., Bras, R.L., and Rybarczyk, S.M. (2001): An Object–Oriented Framework for Hydrologic and Geomorphic Modeling Using Triangulated Irregular Networks. Computers and Geosciences, 27(8): 959–973.

# Part II

# Case studies from the Rhine river catchment and Central Europe

# Large to Medium-Scale Sediment Budget Models - the Alpenrhein as a Case Study

Matthias Hinderer

Institut für Angewandte Geowissenschaften, Technische Universität Darmstadt, Schittspahnstr. 9, 64287 Darmstadt, Germany

**Abstract.** Based on a sediment budget of the Alpenrhein valley and Lake Constance for the Late Glacial and the Holocene, a concept of large-scale sediment budgets for the entire drainage basin of the Rhine River is presented. In particular the following approaches are recommended: (1) Study of sediment volumes in representative quasi-closed subsystems or quantification of losses, (2) upscaling by defining uniform subcatchments using GIS, (3) tracing individual events in floodplain sediments which can be correlated throughout large parts of the river system to secure stratigraphic correlations, and (4) finally calculation of a simplified total budget for the entire system (e.g. Holocene) using approaches (1) to (3). Despite many uncertainties in establishing sediment budgets for open systems, sediment budgets provide the only "hard" data to validate computer simulations of sediment fluxes in the past.

## 1 Introduction

Modern sediment fluxes are usually estimated by measurements of the sediment load of rivers (e.g. Milliman and Syvitski, 1992; Walling and Webb, 1996). Such direct measurements date back not more than to the beginning of the 20th century. To estimate sediment fluxes in former periods, only indirect methods can be applied. One fundamental concept of these estimates is based on the understanding of the present–day erosional and sediment transport processes and their extrapolation to changing framework conditions e.g. climate and land use. This approach became quite popular especially among modellers who simulate palaeo–sediment fluxes based on algorithm validated for present–day conditions (see this issue). In most of these case studies, the calculated palaeo–fluxes are not validated by field data. Therefore, a second fundamental approach is inevitable, which uses the accumulated products of former erosion as archives and converts sediment volumes to sediment fluxes (e.g. Jordan and Slaymaker, 1991; Phillips, 1991; Einsele et al., 1996; Müller, 1999; Sommerfield and Nittrouer, 1999). The basic principle behind this approach is mass conservation in a closed system. This paper discusses the potential as well as limits of sediment budget models and presents a large–scale and long–term sediment budget for the Alpenrhein.

## 2   Theoretical background

Figure 1 demonstrates the basic concept of quantifying sedimentary source–sink systems which are further termed as denudation sediment–accumulation systems (DA systems, Einsele and Hinderer, 1998). Denudation rates are linked with sediment accumulation rates via the ratio drainage area/basin area (Figure 1; for further elementary equations see Einsele and Hinderer, 1998). The quantitative aspect of such DA systems can be approached from two sides: (1) the processes operating in the denudation area are used to estimate sediment accumulation in the sink area; or (2) sediment parameters are applied to reconstruct denudation rates in the drainage area. Such DA systems are subject to complex superposition, interaction and feed back mechanisms of different controlling factors. On a geological time scale of millions of years, uplift and subsidence are the most important controls of DA systems (Figure 1). Considering short–term periods in the order of some thousands to hundred thousand years, uplift and subsidence can be assumed to be in steady–state and variations in climate and land use become dominant.

The idea to determine former sediment fluxes from sedimentation rates and sediment volumes exists since many decades. However, wide–spread application of the sediment budget approach is hampered by (1) the restriction to closed or semi–closed systems, (2) the need of 3D data to estimate sediment volumes, and (3) highly resolved stratigraphic records. Commonly, fluvial environments represent open systems, because part of the eroded sediments are transported to the oceans where they are further dispersed. However, fluvial subsystems such as alluvial fans or pediments may be used to make a minimum estimate of sediment fluxes (e.g. Griffiths and McSaveney, 1986; Campbell, 1998). Detailed sediment budgets for micro–catchments or single gullies have been undertaken by several authors (e.g. Bork et al., 1998). Alluvial fans can be only used for a rough estimate, because stratigraphic markers are rare in coarse–grained sediments and their architecture is highly complex and variable (Mather, 2000). Hence, studies on sediment budgets of alluvial fans or pediments should be coupled with studies on the eroded rock volume in the source area, e.g. reconstruction of former land surfaces (e.g. Ibekken and Schleyer, 1991).

The only total sinks of sediments on land are lakes or endorrheic basins. Lakes are very useful to reconstruct sediment fluxes from sedimentation rates and rates of infill, because they represent sedimentologically closed systems with a close linkage to the denuding hinterland, and – in contrast to fluvial deposits – they exhibit complete sedimentary records. In particular, artificial reservoirs have been successfully used to reconstruct the impact of land use changes because the onset of infilling can be exactly dated by the year in which the reservoir has been put into operation (e.g. Dearing and Foster, 1987; Garcia and Vignoli, 1988; Dearing and Foster, 1993; Duck and McManus, 1994). In small natural lakes, soil erosion may be reconstructed from sedimentation rates (e.g. Foster et al., 1988; Desloges, 1994). In case of laminated lake sediments, highly resolved records of terrigenous sediment input can be achieved (e.g. Zolitschka, 1998). However, such studies represent only local records and their regional relevance should be considered with caution, especially when reconstructing absolute changes. Einsele and Hinderer (1998) presented a general concept of lakes as denudation–accumulation systems which they applied to the large lakes of the world, e.g. Lake Constance (Hinderer and Einsele, 2001). Here, this large–scale concept is applied to the system Alpenrhein/Lake Constance. In addition, further development of this concept is outlined in this paper to improve time–resolution and the treatment of open systems.

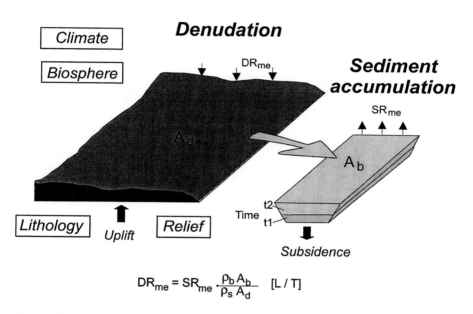

$$DR_{me} = SR_{me} \cdot \frac{\rho_b A_b}{\rho_s A_d} \quad [L / T]$$

**Fig. 1.** Basic concept of denudation–sediment accumulation systems and controlling parameters. $A_d$ = drainage area, $A_b$ = basin area, $DR_{me}$ = mechanical denudation rate, $SR_{me}$ = mechanical sedimentation rate, $\rho_b$ = bulk density of sediments, $\rho_s$ = density of solid rock.

## 3   Late Quaternary sediment budget of the Alpenrhein

### 3.1   General

At present, the Alpenrhein flows into Lake Constance, one of the largest perialpine lakes in Europe (area = $539\,km^2$, volume $49\,km^3$, maximum depth = $252\,m$). The drainage basin of the Alpenrhein and its tributaries covers an area of $7068\,km^2$ and its geology comprises different kinds of lithology related to various tectonic units of the Alps (Figure 2). The headwaters drain the central crest of the Alps. Downstream, the Alpenrhein cuts through the External and Helvetian Alps with major tributaries from the east (Landquart, Ill).

The deep basin of the Alpenrhein valley and Lake Constance has been scoured by repeated advances of the Rhine glacier mainly during the Late Pleistocene (Keller, 1994). The present–day lake was formed at about 17 kyr BP when the Rhine glacier retreated from the foreland after the Last Glacial Maximum (LGM, "Würmian"). The southern end of the initial lake reached far into the present Alpenrhein valley up to Chur. Since this time, the lake has acted as a sediment trap for all sediments transported by the Alpenrhein and its tributaries, i.e. it became sedimentologically closed. Hence, the system Alpenrhein/Lake Constance provides an excellent example to quantify Late Pleistocene to Holocene denudation and infill rates, respectively (Figure 3).

## 3.2 Method

Based on seismic transects, drillings, and morphometry of the valley the volume of the post–LGM infill of the Alpenrhein valley has been calculated (for details see Hinderer, 2001). The maximum thickness of the unconsolidated valley infill amounts to c. 660 m with a minimum bedrock surface at 180 m below sea level. Seismic transects show a single filling cycle with unstructured rocks at the base, interpreted as till deposits, a thick sequence with layered deposits, interpreted as lacustrine and deltaic deposits, and weakly structured alluvial deposits, which are well known from shallow drillings (Eberle, 1987; Pfiffner et al., 1997). Because of this single sequence, pre–Würmian deposits must have been largely removed during the last advance of the Rhein glacier implying sedimentologically open conditions during the LGM.

The $^{14}$C datings of organic matter in the mainly coarse–grained sediments have been used to separate Holocene from Late Pleistocene sediments. sediment volumes have been converted into sediment fluxes by an average density of $2\,g/cm^3$ for Late Pleistocene deposits and $1.7\,g/cm^3$ for Holocene deposits. The higher density of Late Pleistocene deposits is caused by their stronger compacting (Müller, 1995). The absolute timing is based on a regional calibration of $^{14}$C ages to calendar ages (Stuiver et al., 1991).

## 3.3 Results and discussion

The results of the Late Pleistocene to Holocene sediment budget are shown in Figure 4 b,c and compared with the modern sediment yield of the Alpenrhein (Figure 4 a). According to the sediment budget, the Late Pleistocene sediment flux was c. 10–fold higher than at present. This order of magnitude corresponds well with results from other drainage basins in the Alps as well as similar basins in the Northern Hemisphere (Hinderer, 2001). The Holocene sediment flux as calculated from sediment volumes is close to the present sediment flux as recorded from 80 years river load measurements and delta growth rates (Meixner, 1991; Figure 4 a), i.e. the present sediment flux of the Alpenrhein seems to be representative for the Holocene. However, this does not mean that present processes of

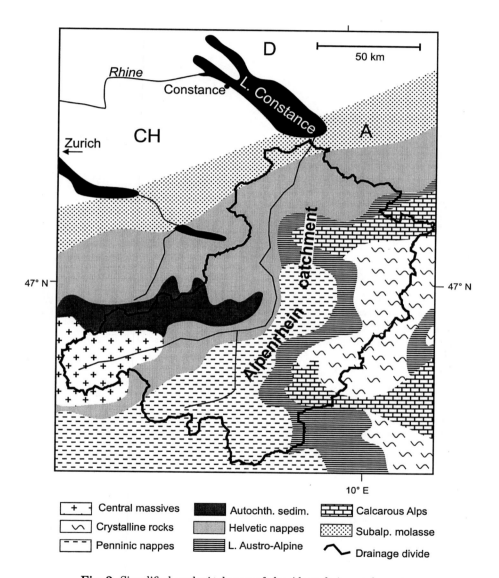

**Fig. 2.** Simplified geological map of the Alpenrhein catchment.

Late Pleistocene to Holocene sediment budget of the Alpenrhein basin

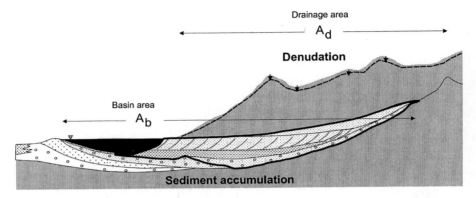

**Fig. 3.** The Alpenrhein catchment and Lake Constance as a closed denudation–accumulation system. The accumulated sediment volume since closure after the LGM is outlined by a bold line and corresponds to a specific lowering of the surface of the catchment (dotted line). Both, denudation rate and accumulation rate during this time are linked by the ratio denudation to accumulation area and the rock density (see text); $A_d$ = drainage area, $A_b$ = basin area.

erosion and sediment transport have remained unchanged over several thousand years. Human impact clearly affects the present sediment flux of Alpine river basins. Thereby, two kind of human impacts exist which have opposite consequences: (1) increase of sediment flux caused e.g. by forest clearings and river short–cuts and (2) decrease of sediment fluxes caused by gravel mining and construction of large reservoirs. The bedload transport of the Alpenrhein is clearly not in steady–state and largely controlled by human impact (Zarn and Oplatka, 1992). The suspended load which provides more than 90% of the total sediment transport seems to be much less affected by direct impacts but no quantitative studies exist to this regard. To unravel man's role in the present sediment flux of the Alpenrhein, a higher resolution of the Holocene sediment budget is required (see section 4).

The results of the Late Pleistocene to Holocene sediment budget can be put into an evolutionary diagram for the Lake Constance basin (Figure 5). Here, the basin volume is expressed as linear accumulation space (volume/accumulation area) which is limited by the lake level. The slight decline of the lake level over time is caused by incision of the rock threshold at the outflow at Constance. The additional accumulation space which is necessary in order to keep a specific river gradient when the lake would be filled and transformed into an alluvial plain is not considered here but would prolongate the life time of the lake. The diagram shows the rapid infill of the basin during the first 5 kyr which corresponds to high denudation rates. According to studies of e.g. Karl and Danz (1969) and Fahn (1978) the mechanical denudation rate in Alpine catchments during the

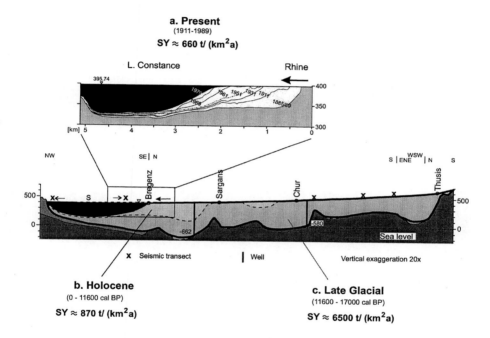

**a. Present**
(1911-1989)

SY ≈ 660 t/ (km²a)

**Fig. 4.** Sediment volumes accumulated in the Alpenrhein valley for various time slices. (a) Present (based on surveys of delta growth), (b) Holocene, (c) Late Pleistocene. The sediment budget of (b) and (c) is based on seismic transects and drillings. Conversion from volumes to sediment yield see text. Modified after Hinderer (2001).

early Holocene was half that during the last 5 kyr. This ratio has been used to divide the Holocene infill rate in the diagram into two sections. Any scenario can be put forward to speculate about the further infilling of Lake Constance. A steady–state scenario would lead to a life time of c. 17 kyr. The life time increases to c. 23 kyr if the additional accumulation space in order to keep the river gradient is taken into account.

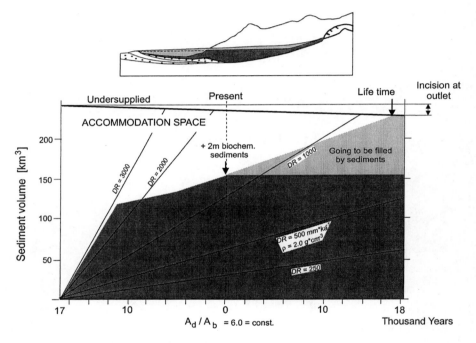

**Fig. 5.** Temporal infill of the Alpenrhein valley and Lake Constance since the Late Glacial und hypothetical life time of the lake in the future for steady–state conditions. Thin lines show the increase of mean sediment thickness in the basin for various denudation rates of the Alpenrhein catchment if the sediment density is $2.0\,\mathrm{g/cm^3}$. $A_d =$ drainage area, $A_b =$ basin area.

## 4 Reading temporal–spatial process dynamics from the sediment budget

Sediments of Lake Constance contain a continuous high resolution record of environmental changes since the Late Pleistocene (Wessels, 1995; Wessels, 1998; Figure 6a). In particular, lake sediments from the northern slope of the Lake

Constance basin are dominated by the interflow of the Alpenrhein and record changes in the sediment flux since the Late Glacial. Based on AMS $^{14}$C, varve counting, and palaeomagnetic methods, Wessels (1995) was able to reconstruct changes of sediment supply of the Alpenrhein since the Late Pleistocene. At the beginning of the Holocene the sediment supply rapidly decreased and lake sedimentation was dominated by authigenic calcite production. After 6 kyr, the detrital sediment supply by the Alpenrhein increased and the authigenic carbonates produced in the lake were diluted. This shift is related to the disappearance of large lakes in the Alpenrhein valley which were dammed by alluvial fans or landslides and subsequently filled up. After 5 kyr BP, intervals of strongly increased sediment input at 4100, 3500 and 2600 yr BP and during the "Little Ice Age" are indicated by increased mean lamina thickness. Despite a deterioration of climate (e.g. Furrer et al., 1987; Schwalb et al., 1994; Wessels, 1995), pollen analyses and archeological evidence show the beginning of human impact at this time (Rösch, 1992; Mayer, 1999).

These temporal changes in the sediment flux of the Alpenrhein have not been quantified yet. For a highly resolved estimate of the variations in time in response to Holocene environmental changes, the delta sediments must be considered as well, because they contain by far the largest sediment volumes. In contrast to the lake area, however, wells and age control of the thick coarse grained delta sediments are scarce so far. The different distribution patterns of detrital sediment in the present lake for typical events or time periods which can be easily derived from numerous sediment cores may provide an indirect method to correlate lake sediments with delta sediments and alluvial sediments of the Alpenrhein (Figure 6b).

Sediment storage and sediment flux are closely related in a temporal–spatial framework. Figure 7 gives a schematic display of this relationship for the system Alpenrhein/Lake Constance since the Late Glacial. At glacial times sediments were stored in the source area and the periglacial foreland (e.g. till deposits). The sediment flux was mainly controlled by the transport capacity of the glacier. A cold and dry climate prevented a significant outflux of sediments from the foreland to the main distributory rivers. This strongly changed during the Late Glacial melting period when the fluvial and glaciofluvial transport capacities rose tremendously. Sediment fluxes reached a maximum. Due to the formation of Lake Constance, however, the sediment flux from the Alpenrhein was decoupled from the foreland. In the headwaters and the foreland a relatively high sediment flux continued whereas it became practically zero in the Alpenrhein valley where sediments were largely deposited. The present situation approaches increasingly steady state conditions but represent still some non–steady state features. The sediment sinks within the headwaters have been successively emptied and sediment flux has declined (Figure 8). The headwaters of Alpenrhein show a relatively low sediment flux which reflects only small volumes of stored sediment at present. Highest sediment fluxes are directed to the main depocenter and largely transferred to the terminal sink of Lake Constance.

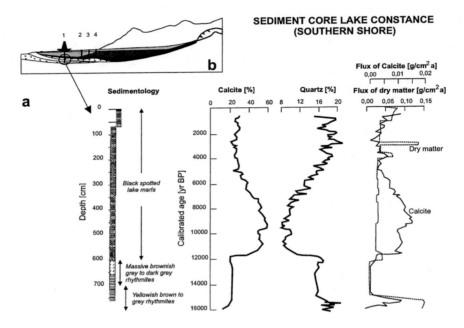

**Fig. 6.** Sediment core from the northern shore of the central Lake Constance (c. position 1 in sketch a, Friedrichshafener Bucht) representing the last 14 kyr. Lithology, mineralogy and fluxes after Wessels (1995). Similar data from more proximal locations (e.g. 2 to 4) are necessary to make an estimate of sediment volumes.

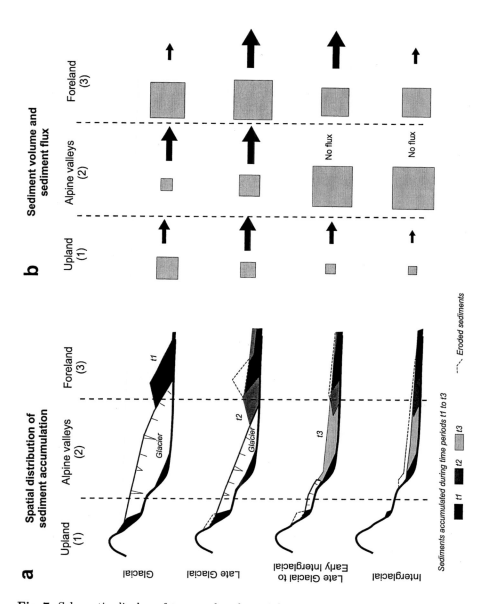

**Fig. 7.** Schematic display of temporal and spatial variation of sediment budget and sediment fluxes at the glacial to interglacial transition according to studies in Alpine drainage basins.

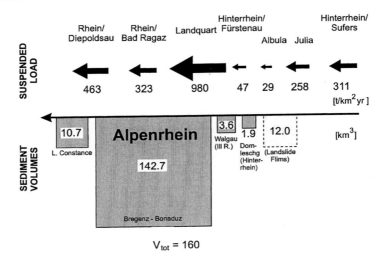

**Fig. 8.** Spatial distribution of sediment volumes in the Alpenrhein valley and Lake Constance basin which have been accumulated since LGM and present sediment yield at various locations within the drainage basin.

Under interglacial conditions, the sediment flux in the foreland is more affected by a glacial time lag because the landscape has still a widespread cover of unconsolidated glacial and glaciofluvial sediments which are eroded on hillslopes or by river incision. Quasi steady–state conditions establish not before a complete infill of the Lake Constance basin and complete removal of glacial sediments. However, because of decreased interglacial process rates, such a new stage of quasi–equilibrium will not be reached before a presumably new glacial cycle with a cyclicity of c. 100 kyr. Pure steady–state conditions would be fulfilled if no further changes in sediment storage occur. This is rarely realised in natural systems.

Figure 9 schematically shows the temporal and spatial time lags of sediment fluxes by plotting the sediment flux as a function of time and distance. Sediment flux versus time shows a systematic increase in amplitude and time lag from headwaters to the foreland. This pattern was termed the "paraglacial cycle" by Church and Ryder (1972). Under equilibrium conditions, the sediment flux continuously declines with increasing distance from the upland to the foreland. This pattern has been observed in many studies and is assigned to the increased storage of sediment, often expressed as a decreasing sediment delivery ratio (e.g. Walling, 1983; Klaghoffer et al., 1992). Conditions of disequilibrium are characterised by deviations from this exponential curve. In Figure 9 two examples are given for glacial and early interglacial conditions, however, other transitional patterns may develop.

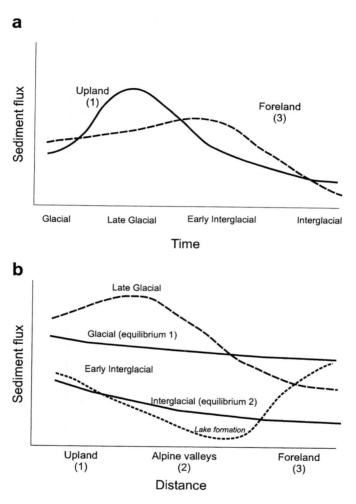

**Fig. 9.** Schematic curves of temporal and spatial variation of sediment flux at the glacial to interglacial transition according to the paraglacial cycle of Church and Ryder (1972) (a) and own studies in Alpine drainage basins (b).

## 5  Concept for the sediment budget approach in an open system

So far, we dealt with relatively simple large–scale sediment budgets in a largely closed system. Most river systems do not have either sedimentologically closed conditions nor they have a continuous sedimentary record of former sediment fluxes. Nevertheless, the fundamental principle of non–steady–state systems as described above is also valid for open system. In the Late Quaternary sediment budget of the Alpenrhein, the disturbance which shifted the system towards non–steady state conditions was of climatic origin. For the Holocene such disturbances have been mainly caused by man, however, superposition or even coupling with climate disturbances are likely.

Figure 10 transfers the schematic display of temporal–spatial changes for disturbed systems from a glacial origin to a human origin. Human disturbance of a hillslope by land use changes, (e.g. clearfelling) subject the soils and regoliths to accelerated erosion and sediment flux. In the schematic display of Figure 10 to the right soils and regoliths are considered as a reservoir of rapidly removable sediments. As shown by many studies, this material is largely re–accumulated in the alluvial plain and the riverine sediment flux increases to a much lower degree than hillslope erosion does (e.g. Trimble, 1975; Walling, 1983). This discrepancy increases in the downstream direction of the system. In a recovery stage of adapted agricultural practices and stabilisation of the alluvial plain, the sediment fluxes decrease but still remain on a higher level than before land use change started. This schematic pattern converts to similar temporal–spatial curves of sediment flux as shown for glacial disturbances in Figure 9 (see also Osterkamp and Toy, 1997).

Figure 11 presents a sketch how small to large–scale sediment budgets together with concepts of regionalisation or upscaling might be used to establish a sediment budget for large and sedimentologically open river basins. The fundamental questions (i) to which extent human impact (or any other disturbance) affected upland erosion and (ii) to which extent the system's response was transferred to the river network can only be answered by local case studies on sediment budgets (distributed approach; e.g. Ferro, 1997). To reach the highest possible accuracy, estimation of accumulated sediment volumes should be compared with lost volumes of material since the disturbance took place (-V1). Because of the open nature of the system, the accumulated sediment volume (+V1') must be less than the eroded volume. The difference correspond to the loss of fines out of the system (+V1"). The shift of the grain–size distribution from weathered to accumulated material may provide an independent method to estimate the volume of lost fines (Atkinson, 1995). The results might be the control for hindcasts from hillslope models.

In a second step, representative regions in the river basin are identified and individual sediment budgets from case studies are extrapolated to a generalised sediment budget of all headwaters. Nowadays, the most appropriate tools with this respect are Geo–Information Systems (e.g. Kothyari and Jain, 1997; Molnar and Julien, 1998). The GIS must include information layers on all parameters

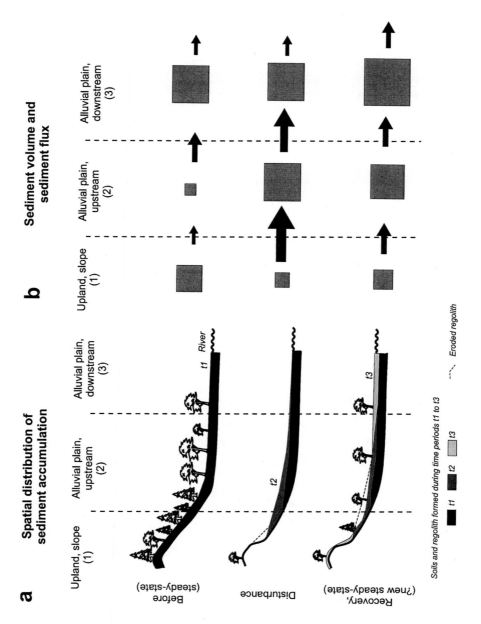

**Fig. 10.** Schematic display of temporal and spatial variation of sediment budget and sediment fluxes due to land use change by man.

relevant for sediment fluxes, e.g. topography, soils, lithology, climate, landuse etc.

In a third step, the transfer of sediments through the river network is investigated. Empirical approaches such as sediment rating curves or computer models can be used to model the downstream transport of the sediments delivered from the headwaters (results of steps one and two). A sediment budget of downstream regions of a basin is hampered by several reasons as there are: (i) increase of sediment dispersal over large areas, (ii) increase of the number of gaps in sedimentary record, and (iii) increase of stratigraphic uncertainties. Hence, such a sediment budget for downstream areas of large river basins can only be established in a very simplified manner, both in terms of space and time. Because of the strong heterogeneity and wide distribution of alluvial deposits an adequate data base would require hundred thousands or more of drilling data which are nowadays only available for few regions (e.g. Berendsen and Stouthamer, 2000, 2001, for the Lower Rhine). Until no comprehensive data base exists, representative localities of alluvial deposits and 2D valley transects must be studied in detail and then extrapolated by considering the large scale fluvial architecture. Summing up all the sediments of the alluvial plain in the time period of interest $(V1"n)$ gives a minimum estimate of the sediment volume removed from the hillslopes in the drainage basin $(-V1)$, because the system is open towards the sea which represent the ultimate sink for the sediment routing system.

## 6    Discussion and conclusions

This paper has presented a Late Pleistocene and Holocene sediment budget of the Alpenrhein/Lake Constance system. Processes of changing weathering regime, transport media, and transport capacity during this time are well reflected in growth and decay of sediment sinks and temporal changes of sediment fluxes. However, this system represents a relatively simple large–scale sediment budget in a largely closed system. To apply the concept of large to medium scale sediment budgets to sedimentologically open systems the following approaches are proposed:

1. Study of sediment volumes in representative quasi–closed subsystems or quantification of losses
2. Upscaling by defining uniform subcatchments using GIS
3. Tracing of individual events in floodplain sediments which can be correlated throughout large parts of the river system to secure stratigraphic correlations
4. Final calculation of a simplified total budget for the entire system (e.g. Holocene) using approaches (1) to (3).

These approaches should be tackled in combination. Approach (1) may be used as a framework and control of regionalised estimates from temporarily better resolved subsystems. Approach (3) needs accurate local studies of sediment budgets. Losses from open systems should be estimated from two sides, i.e. the loss of fines as estimated from the shift of grain size distribution in weathered

## 1) Representative case studies in headwaters

Small to medium-scale sediment budget.
Comparison with models on hillslope erosion

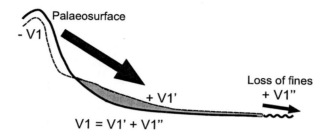

## 2) Upscaling

GIS with layers on topography, landuse,
soils, lithology, climate

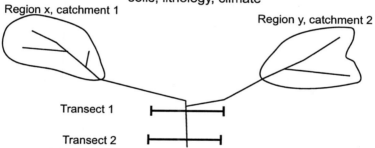

## 3) Entire river basin

Large-scale budget.
Comparison with sediment transport models

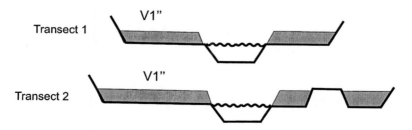

**Fig. 11.** Basic concept of an integrated sediment budget of large drainage basins (further explanation see text).

and accumulated material, and excavated volume of material by the reconstruction of former landsurfaces. (1) may be used as a control in order to calibrate the algorithms and tools of regionalisation (approach 2). Approach (3) might offer the opportunity to estimate sediment volumes that are mobilised by an individual extreme event for which stratigraphic identification is possible over large distances and by accelerated erosion and transport rates of post–event processes. It is also of high socio–economic relevance to get estimates of magnitudes of maximum possible events and their impacts. For longer time periods, e.g. the Holocene, temporal changes of the frequency of events may be identified when analysing the entire sedimentary record.

Despite many uncertainties in establishing sediment budgets for open systems, sediment budgets provide the only "hard" data to validate computer simulations of sediment fluxes in the past. Although several computer codes exist to simulate long–term sediment fluxes and their applicability have been demonstrated in case studies, we can never be sure of their results without controlling the output with the evidence in the field. Therefore, both approaches, modelling and field studies on sediment budgets should be pursued in an interactive manner.

# References

Atkinson, E. (1995): Methods for assessing sediment delivery in river systems. Hydrological Sciences Journal 40(2): 273–280.

Berendsen, H.J.A. and Stouthamer, E. (2000): Late Weichselian and Holocene palaeogeography of the Rhine-Meuse delta. Palaeogeography, Palaeoclimatology, Palaeoecology 161: 311–335.

Berendsen, H.J.A., and Stouthamer, E. (2001): Palaeogeography development of the Rhine–Meuse delta, The Netherlands. Van Gorcum, Assen.

Bork, H.–J., Bork, H., Dalchow, C., Faust, B., Piorr, H.–P., and Schatz, Th. (1998): Landschaftsentwicklung in Mitteleuropa. Klett–Perthes, 328 pp.

Campbell, C. (1998): Postglacial evolution of a fine–grained alluvial fan in the northern Great Plains, Canada. Palaeogeography, Palaeoclimatology, Palaeoecology 139: 233–249.

Church, M., and Ryder, J. (1972): Paraglacial sedimentation: a consideration of fluvial processes conditioned by glaciation. Bull. Geol. Soc. America 83: 3059–3071.

Dearing, J.A., and Foster, I.D.L. (1987): Limnic sediments used to reconstruct sediment yields and sources in the English Midlands since 1765. In: Gardiner, V. (ed.): International Geomorphology. John Wiley & Sons, Chichester. pp. 853–868.

Dearing, J.A., and Foster, I.D.L. (1993): Lake Sediments and Geomorphological Processes: Some Thoughts. In: McManus, J., and Duck, R.W. (eds.): Geomorphology and Sedimentology of Lakes and Reservoirs. John Wiley & Sons, Chichester New York Brisbane Toronto Singapore, pp. 5–14.

Desloges, J.R. (1994): Varve deposition and the sediment yield record at three small lakes of the southern Canadian Cordillera. Arctic and Alpine Research 26(2): 130–140.

Duck, R.W., and McManus, J. (1994): A long–term estimate of bedload and suspended sediment yield derived from reservoir deposits. Journal of Hydrology 159: 365–373.

Eberle, M. (1987): Zur Lockergesteinsfüllung des St. Galler und Liechtensteiner Rheintales. Eclogae geol. Helv. 80(1): 193–206.

Einsele, G., and Hinderer, M. (1998): Quantifying denudation and sediment-accumulation systems (open and closed lakes): basic concepts and first results. Palaeogeography, Palaeoclimatology, Palaeoecology 140: 7–21.

Einsele, G., Ratschbacher, L., and Wetzel, A. (1996): The Himalaya–Bengal Fan denudation–accumulation system during the past 20 Ma. The Journal of Geology 104: 163–184.

Fahn, H.J. (1978): Die Sedimentschüttung in die bayrischen Voralpenseen seit dem Schwinden des Würmeises (zugleich als Mass der Abtragungsleistung in den Flusseinzugsgebieten). Dissertation Thesis, Julius–Maximilians–Universität, Würzburg, 137 pp.

Ferro, V. (1997): Further remarks on a distributed approach to sediment delivery. Hydrological Sciences Journal 42(5): 633–647.

Foster, I.D.L., Dearing, J.A., and Grew, R. (1988): Lake–catchments: an evaluation of their contribution to studies of sediment yield and delivery processes. IAHS Publication No. 174, Sediment Budgets: 413–424.

Furrer, G., Burga, C., Gamper, M., Holzhauser, H.–M., and Maisch, M. (1987): Zur Gletscher–, Vegetations– und Klimageschichte der Schweiz seit der Späteiszeit. Geographica Helvetica 1097(2): 61–91.

Garcia, E.P., and Vignoli, F.O. (1988): Average long–term sediment discharge investigations based on reservoir resurvey data and sediment yield rate factors. IAHS Publication No. 174, Sediment Budgets: 425–429.

Griffiths, G.A. and McSaveney, M.J. (1986): Sedimentation and river containment on Waitangitaona alluvial fan – South Westland, New Zealand. Z. Geomorph. N.F. 30(2): 215–230.

Hinderer, M., 2001. Late Quaternary denudation of the Alps, valley and lake filling, and modern river loads. Geodinamica Acta 14(3): 231–263.

Hinderer, M., and Einsele, G. (2001): The world's large lake basins as denudation–accumulation systems and implications for their lifetimes. Journal of Paleolimnology 26: 355–372.

Ibekken, H., and Schleyer, R. (1991): Source and sediment – A case study of Provenance and mass balance at an active plate margin (Calabria, Southern Italy). Springer–Verlag, Berlin Heidelberg, 286 pp.

Jordan, P., and Slaymaker, O. (1991): Holocene sediment production in Lillooet River basin, British Columbia: A sediment budget approach. Géographie physique et Quaternaire 45(1): 45–57.

Karl, J., and Danz, W. (1969): Der Einfluss des Menschen auf die Erosion im Bergland dargestellt an Beispielen im bayrischen Alpengebiet. Schriftenreihe der Bayrischen Landesstelle für Gewässerkunde 1: 98.

Keller, O. (1994): Entstehung und Entwicklung des Bodensees – Ein geologischer Lebenslauf. In: Holenstein, J., Keller, O., Maurer, H., Widmer, R., and Züllig, H. (eds.): Umweltwandel am Bodensee. Fachverlag für Wissenschaft und Studium GmbH, St. Gallen, pp. 33–91.

Klaghoffer, E., Summer, W., and Villeneuve, J.P. (1992): Some remarks on the determination of the sediment delivery ratio. IAHS Publication No. 209, Erosion, debris flow and environment in mountain regions: 113–118.

Kothyari, U.C., and Jain, S.K. (1997): Sediment yield estimation using GIS. Hydrological Sciences Journal 42(6): 833–843.

Mather, A.E.; Harvey, A. M., and Stokes, M. (2000): Quantifying long–term catchment changes of alluvial fan systems. GSA Bulletin 112(No. 12): 1825 – 1833.

Mayer, B., and Schwark, L. (1999): A 15,000-year stable isotope record from sediments of Lake Steisslingen, Southwest Germany. Chemical Geology 161: 315–337.

Meixner, H. (1991): Das Rheindelta im Bodensee – Seegrundaufnahme vom Jahre 1989. Unpublished report: 1–18.

Milliman, J.D., and Syvitski, J.P.M. (1992): Geomorphic/tectonic control of sediment discharge to the ocean: The importance of small mountainous rivers. The Journal of Geology 100: 525–544.

Molnar, D.K., and Julien, P.Y. (1998): Estimation of upland erosion using GIS. Computer & Geosciences 24(2): 183–192.

Müller, B.U. (1995): Das Walensee-/Seeztal – eine Typusregion alpiner Talgenese, Ph.D. thesis, Universität Bern, Bern, 219 pp.

Müller, B.U. (1999): Paraglacial sedimentation and denudation processes in Alpine valleys of Switzerland - An approach to the quantification of sediment budgets. Geodinamica Acta 12: 291–301.

Osterkamp, W.R., and Toy, T.J., 1997. Geomorphic considerations for erosion prediction. Environmental Geology 29(3–4): 152–157.

Pfiffner, O.A., Heitzmann, P., Lehner, P, Frei, W., Pugin, A., and Felber, M. (1997): Incision and backfilling of Alpine valleys: Pliocene, Pleistocene and Holocene processes. In: Pfiffner, O.A. (ed.): Deep structure of the Swiss Alps and results of NRP 20. Birkhäuser–Verlag, Basel, Boston, Berlin, pp. 265–288.

Phillips, J.D. (1991): Fluvial sediment budgets in the North Carolina Piedmont. Geomorphology 4: 231–241.

Rösch, M. (1992): Human impact as registered in the pollen record: some results from the western Lake Constance region, Southern Germany. Vegetation History and Archaeobotany 1: 101–109.

Schwalb, A., Lister, G.S., and Kelts, K. (1994): Ostracode carbonate $\delta^{18}O-$ and $\delta^{13}C-$ signatures of hydrological and climatic changes affecting Lake Neuchâtel, Switzerland, since the latest Pleistocene. Journal of Paleolimnology 11: 3–17.

Sommerfield, C.K., and Nittrouer, C.A. (1999): Modern accumulation rates and sediment budget for the Eel shelf: a flood–dominated depositional environment. Marine Geology 154: 227–241.

Stuiver, M., Braziunas, T.F., Becker, B., and Kromer, B. (1991): Climatic, solar, oceanic, and geomagnetic influences on Late–Glacial and Holocene atmospheric C–14/C–12 change. Quaternary Research 35: 1–24.

Trimble, S.W. (1975): Denudation studies: Can we assume stream steady state? Science 188: 1207–1208.

Walling, D.E. (1983): The sediment delivery problem. Journal of Hydrology 65: 209–237.

Walling, D.E., and Webb, B.W. (1996): Erosion and sediment yield: a global overview. IAHS Publication No. 236, Erosion and sediment yield: global and regional perspectives: 3–20.

Wessels, M. (1995): Bodensee–Sedimente als Abbild von Umweltveränderungen im Spät– und Postglazial. Göttinger Arbeiten zur Geologie und Paläontologie 66: 87.

Wessels, M. (1998): Late–Glacial and postglacial sediments in Lake Constance (Germany) and their palaeolimnological implications. Arch. Hydrobiol. Spec. Issues Advanc. Limnol. 53: 411–449.

Zarn, B., and Oplatka, M. (1992): Die Entwicklung der Rheinsohle in den nächsten 100 Jahren. International Rheinregulierung: 402–404.

Zolitschka, B. (1998): A 14,000 year sediment yield record from western Germany based on annually laminated lake sediments. Geomorphology 22(1): 1–17.

# Impact of Climate and Land Use Change on River Discharge and the Production, Transport and Deposition of Fine Sediment in the Rhine basin – a summary of recent results

Hans Middelkoop[1] and Nathalie E.M. Asselman[2]

[1] Utrecht University, Dept. of Physical Geography, P.O. Box 80.115, 3508 TC Utrecht, the Netherlands
[2] WL — delft hydraulics, P.O. Box 177, 2600 MH Delft, the Netherlands

**Abstract.** Over the past decennia, the palaeogeographic development of the Rhine–Meuse delta in the Netherlands has been extensively studied and an extremely detailed database of the Holocene delta architecture has been established. The delta offers a unique palaeo–environment to study both palaeogeography and avulsions on a timescale of millenia, because of the relatively complete geological record, as a result of rapid aggradation during the Holocene, governed by relative sealevel rise and land subsidence. The palaeogeographic reconstruction provided many new insights in the characteristics of the fluvial system, such as period of existence of individual channel belts, avulsion frequency and the factors influencing these characteristics over time scales of centuries to millennia. In recent years, research attention also became focused on the Rhine basin, which is the main source area of the water and sediment for the delta. These studies considered a decadal to century time scale, and aimed at assessing the potential impacts of future changes in climate and land use on the river Rhine discharge regime as well as on the production of fine sediment and the net transport of wash load from the upstream basin to the lower Rhine delta. For this purpose, a suite of GIS–embedded models has been developed that simulates this sequence of processes. The results indicate that climate change may accelerate erosion rates. However, land use changes, that foresee a reduction in arable land, lead to a net reduction of erosion in the German part of the basin. Due to inefficient sediment delivery, increasing erosion in the Alps has little effect on the sediment load farther downstream. Suspended sediment loads transported into the lower delta are expected to decrease by 13 %. Due to the combined effect of increased peak flows, discharge–dependent sediment transport and trapping efficiencies of floodplains, long–term average sedimentation rates at low floodplain may decrease, while sedimentation rates at high floodplains may accelerate.

## 1   Introduction

Within the framework of LUFICS, the fluvial system is considered by the following suit of components: (1) production of water, sediment, nutrients and pol-

lutants; (2) delivery of these into the river system; (3) their transport through
the river system; (4) sinks and storages within the system and, (5) the delivery
to the mouth of the fluvial system. The Netherlands, situated at the delta of the
river Rhine experiences the net result of all these processes in the upstream river
basin. Over a period of millennia and more, the country owes its existence to the
natural deposition of sediment produced in the upstream basin. With increasing
human influence, the Rhine has become a river that provides a vital source of
water for environment, water management and many economic functions. Ever
since people have built river dikes in the late Middle Ages, flooding has been a
serious threat for people, live–stock, and economy. Deposition of fine sediments
still continues on the embanked floodplains and on the Rhine–Meuse estuary.
During the past 100 years or so, the pollution of the river has become a major concern for health and environment. All these issues will receive continuous
interest in the near future owing to climate change, large–scale landscaping measures for ecological restoration and to reduce flood risks that are foreseen along
the lower Rhine River, economic demands on channel dimensions, and dredging
in the Rotterdam Harbour region, together with ecological development of other
parts of the estuary. All these policy–relevant issues demand a scientific understanding of the processes of the fluvial system. Research will not only concern
the delta, but it also is necessary to have insight in the processes and controls
in the upstream river basin that determine the amounts of water, sediment, nutrients, and pollutants reaching the delta. The present paper gives a summary
of research carried out on the Rhine, related to the LUCIFS–Rhine framework.
After a short introduction on research carried out at Utrecht University on the
Holocene palaeogeography of the Rhine–Meuse delta, the emphasis of the paper
is on current research on the potential impacts of future changes in climate and
land use on the river Rhine discharge regime as well as on the sediment budget
in the river basin.

The objective of the present study is to analyse the sediment delivery and
transport processes at the scale of the entire Rhine basin and to estimate the
potential effects of changes in climate and land use on the production of fine
sediment and the net transport of wash load from the upstream basin to the
lower Rhine delta. For this purpose, a suite of GIS–embedded models has been
established that simulates river discharge and the production, and transport of
wash load through the drainage network and the deposition of overbank fines
along the lower river reaches. The study was carried out within the framework
of the Dutch National Research Programme on Global Air Pollution and Climate Change (NRP) (project no. 952210) and contributes to the research of
the Commission for the Hydrology of the Rhine basin (CHR). The study is described more extensively in Kwadijk (1993), Van Dijk and Kwaad (1999), Van
Dijk (2001), Asselman (1997; 1999) and Middelkoop (1997; 2000).

## 2 Paleogeographic development of the Rhine–Meuse delta

Over the past 35 years, the Holocene Rhine–Meuse delta has been studied extensively (Berendsen and Stouthamer, 2001). The research has focused on detailed mapping of the fluvial geomorphology and geology of the delta, analysis of fluvial sedimentology and facies, chronology of channel belts, avulsion (abandonment of a channel belt by a stream in favour of a new course) history, and factors controlling the palaeogeographic development of the delta (Berendsen and Stouthamer, 2000; Stouthamer and Berendsen, 2000; Stouthamer, 2001).

### 2.1 Research methods

To reconstruct the Holocene palaeogeography, detailed geological and geomorphological mapping (scale 1:10,000) of the delta was carried out. Map units are based on landform, lithology and age (Berendsen, 1982). Lithological information was generally obtained by coring, or, whenever available, from excavations. A total of more than 200,000 borings was carried out, which amounts to a coring density varying from 30 to 350 per km$^2$. Drilling depth was at least 2–3 m, with about 20 % of the boreholes penetrating the entire Holocene sequence (up to 15 m depth). All sediment cores were described in the field at 10-cm intervals. This involved a description of texture, organic material content, gravel content, median grain size, colour, iron and calcium carbonate content, occurrence of groundwater, shells and other characteristics. Occasionally, laboratory checks of field descriptions were carried out.

To reconstruct the palaeogeographic development, 206 erosional remnants of channel belts were described, named and dated. Dating of the channel belts was done using (1) the relative depth of overbank deposits of different channel belts; (2) de–calcification depth of deposits; (3) gradients of surfaces connecting the tops of channel deposits of channel belts; (4) pollen analysis; (5) archaeological artefacts; (6) historical evidence: maps or written documents; (7) $^{14}$C dating of beginning and end of deposition of channel belts. A total of 1200 radiocarbon ages was determined. All channel belts were digitized and stored in a Geographical Information System database (GIS) in Arcinfo format. Palaeogeographic maps can be produced from the GIS database for any moment during the Holocene. The accuracy of $^{14}$C–dates of channel belts is considered to be ± 100 years. The maps can be downloaded from the Internet from: http://www.geog.uu.nl/fg/palaeogeography.

### 2.2 Palaeographic development

During the Early Holocene, rivers were incised in Late Weichselian fluvial terraces, leading to a period of non–deposition, that, at a given location, lasted until the intersection of the Late Weichselian terraces and Holocene deposits had shifted upstream of that location. Sealevel rise is held responsible for the inland shift of the terrace intersection. Aggradation started in the near–coastal area between 9000 and 8000 yr BP, and rapidly shifted inland. Approximately 6000 yr BP this rapid shift slowed down at the western fault of the Peel Horst in the central part of the delta. After 6000 yr BP the terrace intersection rapidly shifted inland again, although the rate of sealevel rise decreased. Figure 1 shows the position and age of the channel belts of the Rhine–Meuse delta. The oldest Holocene channel belts are located along the northern part of the delta. Between approximately 8000 and 4000 yr BP a river pattern developed in the central western part of the delta, consisting of an anastomosing complex of straight and narrow channels, with a low width/thickness ratio of the channel sandbody. After 5500 yr BP the main course of the Lower Rhine developed in the NW part of the delta, and discharged west of Utrecht to the North Sea. This Old Rhine channel remained the main Rhine branch until 1122 AD, when its trunk course was dammed, and its drainage was diverted to the River Lek, that gradually had become more important. After 2000 yr BP, when discharge and sediment load seem to have increased, and coastal erosion occurred in the Maas estuary near Rotterdam, main discharge of the Rhine shifted to the southwestern part of the delta. After 1000 yr BP human influence became increasingly important. Main man–induced changes include the embankment of the rivers, digging of canals, and the construction of groynes and weirs.

The palaeogeographic evolution can be explained as a complex interaction of various factors influencing the river dynamics. These factors are: the shape of the Pleistocene valley, lithological composition of the substrate, sealevel rise, neotectonic movements, coastal evolution, discharge and sediment load variations, and human interference. Especially the influence of neotectonics on Holocene fluvial evolution and avulsion locations in the Rhine–Meuse delta is now supported by convincing evidence, as many avulsions occurred on faults.

The Holocene Rhine–Meuse delta forms the last sink of sediments, before the rivers enter the sea. The research onto the delta has provided a detailed record of the sedimentation history over the past 7500 years. Within the LUCIFS–Rhine project, a next step is to establish a link between sediment deposition in the delta and discharge and sediment yield from the upstream basin. This requires insight in these processes,together with modelling tools at the basin scale, which may enable the determination of sediment yield and the sensitivity to changes in climate and land use. These issues are addressed in the present paper for present–day and future climate conditions.

## 3   River discharge and sediment yield in the Rhine basin

The discharge of the river Rhine is mainly determined by the amount and timing of precipitation, snow storage and snow melt in the Alps, evapotranspiration surplus during the summer period, and changes in the amount of groundwater and soil water storage. The Alpine rivers are governed by a snow melt regime, with a pronounced maximum in summer. This maximum is generated by storage of precipitation in the snow cover in the winter, and its melting in spring and

hectares to a few square kilometres only. It must contain complex geoarchives that allow a detailed quantitative and qualitative reconstruction of soil erosion, land use change and climate change.

2. First field survey
   The first field survey helps to identify the best excavation sites in the deposition areas.

3. Detailed field campaign
   Trenches must be dug at the deposition sites down to the base of the late Pleistocene sediments or the base of the Holocene soils. A grid of trenches and additional drillings must allow the quantification and the interpretation of the development of the whole catchment area.

4. Definition of a process based stratigraphy
   A stratigraphy is defined as a chronological, genetic and spatial order of landscape development, i.e. of erosion and deposition processes, and of soil formation in the catchment area.

5. Dating of sediments and thus soil erosion and related phenomena

6. Laboratory sediment and soil analysis

7. Finalising the process based stratigraphy including all data collected

8. Quantification of soil erosion

9. Detailed and final analysis of written documents

10. Final interpretation of landscape development under human influence

# 3 Methods used at the investigation area Biesdorfer Kehlen

The investigation area was chosen because it contains several geoarchives, colluvial fans of gully systems and deposition sides on the concave hillslopes.

**Field work**

Deep and long trenches on the deposition sites of the catchment, on the concave hillslopes with colluvial deposits, on the colluvial fans of gullies and on gully infillings were dug to provide an exact and complete picture of the deposition site. Additional drillings were made to provide data for the correlation of the layers and soil horizons found in the trenches. The walls of the trenches were cleaned. The layers of sediments, soil formations and other characteristics found on the cleaned walls were marked. For the drawings a horizontal scale of 1:20 and a vertical scale of 1:10 were used. A description of sediment and soil characteristics and a description of the spatial distribution of the features followed (e.g. AG Boden, 1994, and Bork and Dalchow, 2000). The collected data resulted in the first process based stratigraphy.

**Dating methods**

Dendrochronology, archaeological dating of pottery and of other cultural remains, Optical Stimulated Luminescence (OSL) dating of sediments, radiocarbon dating of charcoal and the identification of tracers ($^{137}$Cs and $^{210}$Pb concentrations) were done to date the sedimentation, soil erosion and soil formation phases found in the geoarchives.

The dendrochronological datings were carried out by K.-U. Heussner, DAI, Berlin, and the OSL dating were done by the Rathgen Forschungslabor, Berlin. Radiocarbon dates were established with the AMS method by the Leibniz–Labor, Kiel. The radiocarbon calibration program REV. 3.0 was used for calibrating the radiocarbon ages. U.-K. Schkade, Mo. Frielinghaus, Y. Li, and M. Naumann, measured and interpreted the concentration of the tracers $^{137}$Cs and $^{210}$Pb (Schkade et al., 1999).

## Laboratory work

In the laboratory particle size distribution, bulk density and concentration of organic matter content have been measured at the laboratory of the Institute of Geoecology, University of Potsdam.

## Reconstruction of land use and climate evidence from interviews

Interviews with local residents gave evidence of land use and landscape changes of the last decades. A detailed analysis of contemporary written documents was not possible, because historical documents were not obtainable.

## Definition of a process based stratigraphy and final interpretation

The final interpretation, a synthesis of all data collected, was done for the investigated catchments as well as the definition of a process–based stratigraphy, the chronological and spatial order of processes that shaped a catchment area until the present day.

## Quantifying soil loss

Deposition rates have been calculated by firstly measuring the volumes of the individual sediment layers or sequences, integrating the data collected in all drillings and trenches. Secondly, the periods of deposition were defined by a dated stratigraphy. Then the deposition rates were calculated. The source area of the sediments was reconstructed and measured. Based on the volume and the weight of material deposited and on the size of the catchment area, soil erosion rates in tonnes per hectare per year were calculated. The bulk density used for the calculations was $1.7\,\mathrm{g\,cm^{-3}}$.

# 4   The investigation area Biesdorfer Kehlen

The investigation area is situated 60 km east of Berlin near the town of Wriezen (Figure 1), where a moranic plateau is connected by steep slopes with the alluvial plain of the river Oder. The gully systems, which cut from the alluvial plain into the moranic plateau are described in the literature as being formed in the late Pleistocene (Franz et al., 1970, 38 ff; Marcinek and Nitz, 1973, 230 ff). The chosen investigation area is characterised by two long U–shaped valleys, the Biesdorfer Kehle in the east and the Sandschlenke at the northern part of the investigation area (Figure 2). Several small V–shaped secondary gully systems have cut in from the valleys into the moranic plateau. The aim of our investigations was to find out about the processes that led to the development of two of those secondary gully systems and their catchment areas. Therefore, we divided the investigation area (Figure 2) in three parts hereafter referred to as northern, southern and eastern investigation areas.

**Fig. 1.** Position of the research area Biesdorfer Kehlen in Germany.

In the southern investigation area (Figure 2), on the concave slope, we analysed nine trenches, one of them greater than 70 m in length. The gully system is presently covered by forest. It cut from the valley Biesdorfer Kehle up into the steep and the lower concave parts of the slope, into the arable land of the southern investigation area. One gully–finger of this system was analysed (Figure 2). At the eastern investigation area the colluvial fan which has formed in the U–shaped valley Biesdorfer Kehle was studied (Figure 2).

The northern investigation area (Figures 2 and 3) is characterised by several small gully systems and their colluvial fans. The gully system we analysed has a clearly defined colluvial fan that formed in the U–shaped valley, Sandschlenke. Today the catchment area of this gully system is waste land. We analysed trenches and bore holes on the colluvial fan, in the gully system and in the catchment area.

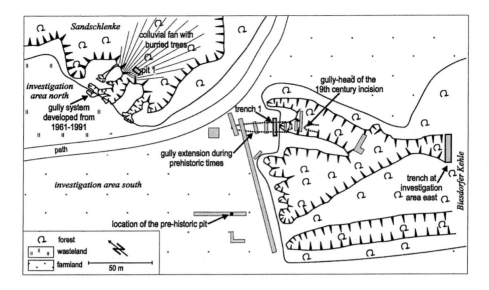

**Fig. 2.** Sketch of the investigation area at the Nature Reserve Biesdorfer Kehlen.

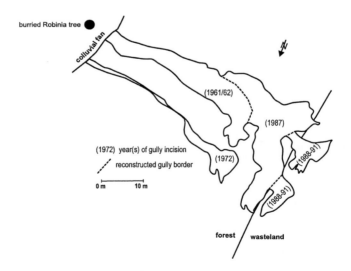

**Fig. 3.** Schematic drawing of the northern gully system, showing the progression of gully erosion.

The analysis of $^{137}$Cs–concentrations was carried out using the sediment record from pit one (Figure 2 on the colluvial fan of the northern investigation area. The dendroecological and dendrochronological investigations were performed on living trees at both investigation areas.

## 5   Description of two trench profiles, from the southern and northern investigation areas

The following profiles are examples for the more than 50 profiles analysed in the whole investigation area. They show the complexity of studied geoarchives.

### Example 1: Trench 1 from the southern investigation area

Trench 1 was dug at the farmland of the southern investigation area, about 2 m south of today's gully head (Figure 2). It revealed a completely filled part of the gully system that is not visible from the recent soil surface any more. The numbers of layers and soil horizons refer to Figure 4. A description of the layers is shown in Table 1.

The first gully incision cut into the late Pleistocene loamy–sandy moranic layers (layer 1) and sand and gravel layers (layer 2). The first gully filling (layer 3) was of loamy–sandy material. Clay enrichment bands were found in this deposition as a result of soil formation.

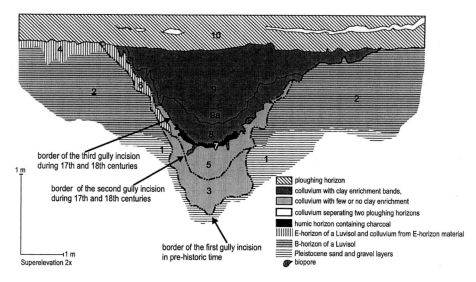

border of the third gully incision during 17th and 18th centuries

border of the second gully incision during 17th and 18th centuries

border of the first gully incision in pre-historic time

1 m

1 m
Superelevation 2x

ploughing horizon
colluvium with clay enrichment bands,
colluvium with few or no clay enrichment
colluvium separating two ploughing horizons
humic horizon containing charcoal
E-horizon of a Luvisol and colluvium from E-horizon material
B-horizon of a Luvisol
Pleistocene sand and gravel layers
biopore

**Fig. 4.** Schematic drawing of trench 1 at the southern investigation area. Today this part of the gully is filled completely and it is used as arable land.

A Luvisol formed in the sediments of layer 2; remains of its E–horizon were found at the southern rim of the filled gully (layer 4). Eroded E–horizon material accumulated in the gully (layer 5). Layer 6 is of loamy–sandy material deposited in the gully. A humic horizon formed, containing charcoal (layer 7). Layers 8, 9 and 9a are loamy–sandy colluvial infillings of the gully. In layer 9 and 9a clay enrichment bands up to 5 cm in thickness were found. The ploughing horizon (layer 10) was divided by a layer of pale sand (layer 11).

**Table 1.** Description of trench 1 at the southern investigation area.

| layer | colour after Munsell (2000) | description | dating (not all samples are taken from this trench but from comparable layers) |
|---|---|---|---|
| 1 | 2,5 YR 4/6 | loamy sand, containing hardly any clay enrichment bands | |
| 2 | 10 YR 4/6 –5/4 | gravel and sandy layers with clay enrichment | |
| 3 | 10 YR 4/4 | loamy sand, colluvium | $2\sigma$ BC 2857–2495 (BP 4080 ± 30, KIA 6019, Leibniz-Labor Kiel) |
| 4 | 10 YR 4/4 | loamy sand, E–horizon | |
| 5 | 10 YR 4/4 | loamy sand, colluvium derived from E–horizon material | |
| 6 | 10 YR 4/4 | loamy sand, colluvium | $2\sigma$ AD 1638–1798, 1943–1954 (BP 250 ± 30, KIA 6021, Leibniz-Labor Kiel) |
| 7 | 10 YR 2/2 | humic horizon | $2\sigma$ cal AD 1674–1778, 1802–1823, 1827–1885, 1912–1942, 1954 (BP 146 ± 20 KIA 6421 Leibniz-Labor, Kiel) |
| 8 | 10 YR 4/3 | loamy sand, colluvium containing clay enrichment bands | |
| 9 | 10 YR 4/4 | loamy sand, colluvium containing clay enrichment bands up to 3 cm thick | |
| 9a | 10 YR 4/4 | loamy sand, colluvium containing clay enrichment bands up to 5 cm thick | |
| 10 | 10 YR 3/2 | loamy sand, ploughing horizon | |
| 11 | 10 YR 3/4 | loamy sand, colluvium, still visible between two ploughing layers | |

**Example 2: Pit 1 on the colluvial fan, northern investigation area**

Figure 5 shows a pit dug at a Robinia tree on the colluvial fan at the northern part of the research area. The trunks of the trees standing on the colluvial fan are buried under up to 120 cm of sediment. To analyse the development of the fan, to grasp its extensiveness and to estimate its thickness, we opened 36 pits at buried trees all over the colluvial fan.

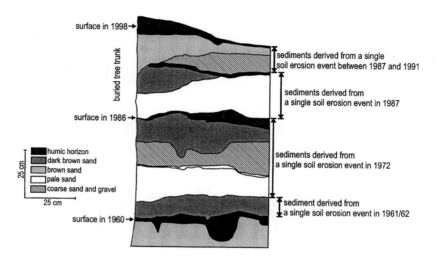

**Fig. 5.** Schematic profile at a buried Robinia tree on the colluvial fan of the northern investigation area (pit 1).

Several layers of colluvial material derived from different sources buried the roots and the trunk of the tree at pit one with one metre of sediment. The pale sands with low or no content of humic material and the gravel layers were derived from gully erosion. The brown and dark brown sands are a mixture of eroded humic horizon material from the catchment area and material from gully erosion. In several layers humic horizons developed (Schmidtchen et al., 2001).

# 6   Landscape changes at Biesdorfer Kehlen

A complex stratigraphy in chronological order was developed for the investigation areas at the nature reserve Biesdorfer Kehlen. Landscape changes due to changing human impact have been reconstructed from Neolithic Times until 1991.

**Phase 1:**

In the Weichselian a loamy–sandy ground moraine was deposited. Under periglacial conditions the U–shaped valleys Biesdorfer Kehle and Sandschlenke were formed. A gelisolifluction layer build up on the lower slopes of the moranic plateau.

## Phase 2:

In the early Holocene soil development under woodland vegetation was the main process. Today this soil is completely eroded.

## Phase 3:

In the late Neolithic the clearing of the natural woodland began a chain of human impacts to the landscape.

A piece of charcoal found in the oldest gully infilling at the southern research area (Figur 4, layer 3) was dated into the late Neolithic (Radiocarbon dating: Two sigma range cal BC 2857–2495; BP 4080 ± 30, KIA 6019, Leibniz–Labor Kiel). Evidence of prehistoric human influence was found on the slope south of the investigated gully–part (Figure 2), where a prehistoric pit filled with charcoal and pottery fragments could be dated into the younger Bronze Age (Radiocarbon dating: Two sigma range cal BC 1185 – 900; BP 2845 ± 25, KIA 7124, Leibniz–Labor Kiel).

In the used prehistoric landscape intensive rainfall and subsequent runoff led to the first human induced gully erosion. The gully refilled slowly afterwards. The remains of this filling contained the Neolithic charcoal mentioned above.

After the Bronze Age pit was built, material from hillslope erosion accumulated on the concave hillslope of the southern investigation area up to 50 cm in thickness. Following these two erosion events, the farmland was abandoned in the catchment area and woodland regenerated again.

## Phase 4:

Phase 4 is characterised by intensive soil formation. A Luvisol developed under woodland. The intensity of the soil formation indicates a long period of geomorphic stability in the catchment area. Dense vegetation cover prevented soil erosion totally. Remains of the Luvisol are still preserved at the lower concave slope of the southern investigation area (Figure 4, layer 4). The leaching horizon developed several dm deep into the sediment. The clay enrichment hardened the moranic material and the periglacial solifluction layer.

Two pieces of charcoal collected from the humic horizon of the Luvisol were dated into the early Medieval Times (Radiocarbon dating: Two sigma cal AD 879 – 1018 and AD 654 – 776; BP 1107 ± 31 and BP 1315 ± 25, KIA 6419 and KIA 7123). The $^{14}$C–dates indicate that the soil formation took place from the Iron Age until the Slavonic settlement in early Medieval Times. Therefore geomorphic stability lasted for a maximum of 3000 years.

**Phase 5:**

Wood clearings and farming started again in early Medieval Times. The land use of the Slavonic people is evident from marks of mouldboard ploughing found in the humic horizon of the Luvisol.

At the northern investigation area material from a humic horizon eroded. The deposition was found at the lower part of the steep margin of the moranic plateau (Figure 2). This sediment was dated twice by the OSL method to AD $1230 \pm 40$ (OSL 1 and 2, Rathgen Forschungslabor, Berlin).

At the southern investigation area runoff on farmed land led to gully erosion in the $17^{th}$ or $18^{th}$ centuries. A second gully cut into the sediments of the prehistoric gully. This incision filled slowly with loamy–sandy material from hillslope erosion, most likely under grassland vegetation (Figure 4 and 6). Charcoal found on the bottom of this infilling was dated into the second half of the $17^{th}$ and the $18^{th}$ centuries (Radiocarbon dating: Two sigma cal AD 1638–1798, 1943–1954; BP $250 \pm 30$, KIA 6021, Leibniz–Labor, Kiel). When sedimentation in the gully ceased, a humic horizon developed in the sediments of the partly filled gully. Several firesides were found on this humic horizon. Charcoal from one fire site could be dated to the $18^{th}$ century (Radiocarbon dating: Two sigma cal AD 1674–1778, 1802–1823, 1827–1885, 1912–1942, 1954; BP $146 \pm 20$, KIA 6421, Leibniz Labor, Kiel). The development of a humic horizon indicates a few years without soil erosion in the catchment area. It is likely that the whole area was used as grassland. After a few years farming recommenced. This facilitated soil erosion in the catchment area. The gully remained under grassland. Sandy–loamy material accumulated in the gully and filled it slowly.

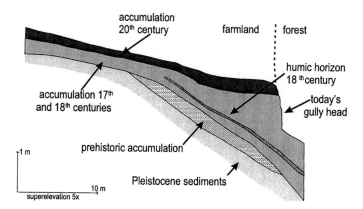

**Fig. 6.** Schematic profile through the gully fillings at the southern investigation area, including sediments accumulated in the gully and colluvium deposited at the concave hillslope.

A third gully incised at the southern investigation area (Figure 2). It eroded the sediments that filled the second gully in parts. A pottery fragment was found in the accumulations of the colluvial fan belonging to the gully system (eastern investigation area). It was dated AD 1855 $\pm$ 32 using the thermoluminescene method (Rathgen Forschungslabor, Berlin) and can be correlated with the third gully incision. Therefore the third gully is likely to have developed in the second half of the 19th century or later.

**Phase 6:**

Dendrochronological investigations of trees that presently grow in the gully system of the southern investigation area indicate that the gully system as we find it today must have been stable for at least 70 years (Dendrochronological datings by K.–U. Heussner, DAI, Berlin). But in the catchment area on the upper slope of the southern investigation area several erosion events took place until today. The accumulations can be found on the concave parts of the slope. They contain many small brick fragments indicating a deposition in the 20th century (Figure 6). Layers of pale sand divide two ploughing layers (Figure 4, layer 11), indicating soil erosion between two ploughings.

In the northern investigation area the first investigated gully erosion occurred in the second half of the 20th century. Trees at the colluvial fan of the gully have partly sediment buried trunks (Figure 3 and 5). Dendroecological investigations on those trees show a reduced ring growth for the years 1961/62 compared with the non–buried trees in the surrounding area. Interviews with local residents revealed that in the first years of the 1960s extreme rainfall and runoff led to gully erosion on the farmed land. This event buried the trees on the colluvial fan with up to 13 cm of sediment. A furrow situated at the lower rim of a field, only a few metres away from the steep slope descending to the valley Sandschlenke, filled with water from high runoff. The concentrated overflow at the lowest part of the furrow caused the gully erosion (Schmidtchen et al., 2001). Parts of the field had to be abandoned following the gully erosion and a new furrow was ploughed further upslope.

The second furrow overflowed after heavy rainfall and subsequent runoff on the field and led to the incision of a new gully. This erosion event can be correlated with one of the heaviest rainfall events measured in Bad Freienwalde, about 10 km north of the investigation area. It rained 66.1 mm on May 20 1972. In the accumulations on the fan of the second gully incision a humic horizon developed, which was still exposed when the Chernobyl fall out reached the area in 1986. Investigations of the $^{137}$Cs content in the sediments of the colluvial fan showed the highest peak in the humic horizon (Schkade et al., 1999, Schmidtchen et al., 2001).

The next gully incision accumulated up to 28 cm of material on the colluvial fan. Two more gully heads cut into the field (Figure 3) due to the concentrated overflow of the third field furrow. This event can be correlated with the heaviest rainfall event in the period from 1969 to 1989, with 89,1 mm on July 18 1987 (measured at Hasselberg, 5 km west of the investigation area). The year following this event, the Robinia trees on the colluvial fan had a reduced growth of rings. A new humic horizon developed in the accumulations from 1987, indicating a few years without accumulation on the colluvial fan.

Again parts of the field were abandoned due to the gully erosion. Once more a new field furrow was ploughed further upslope. A fourth heavy rainfall event caused runoff that filled and flooded the field furrow resulting in the fourth gully incision. This event happened prior to 1991 before the catchment area was turned into waste land, and no more soil erosion occurred.

# 7   Balances of soil loss

For the northern investigation area soil loss derived from hillslope erosion and gully erosion can be calculated from 1961 to 1991. In the colluvial fan of the gully system $1250 \, m^3$ of soil were deposited. The volume of the gully system amounts to $930 \, m^3$. As a result a difference of $320 \, m^3$ from the measured volume of the colluvial fan and the volume of the gully system occurs. This additional material must have been eroded in the catchment area of the gully system, transported through the gully system and deposited on the colluvial fan. The size of the catchment area of the gully system and the size of the erosion area for material derived from hillslope erosion transported through the gully and deposited on the colluvial fan is 1.15 ha.

On the concave slope above the gully system sediments with a volume of $270 \, m^3$ were deposited additionally. Modern artefacts, e.g. pieces of plastic, date those accumulations to the second half of the 20th century. The size of the erosion area for the deposits on the slope is 1.05 ha.

Altogether a total amount of $590 \, m^3$ of material results from hillslope erosion. It is equivalent to $1.8 \, mm \, a^{-1}$ or $30 \, t \, ha^{-1} \, a^{-1}$ in 30 years (Table 2).

**Table 2.** Deposition rates from hillslope erosion at the northern investigation area in 30 years from 1961 until 1991 ($1 \, m^3$ of colluvium is equivalent to $1.7 \, t$ of soil).

| material source area and deposition area | hillslope erosion (in $m^3$) | gully erosion (in $m^3$) | hillslope erosion (in $t \, ha^{-1} \, a^{-1}$) |
|---|---|---|---|
| material from hillslope erosion deposited on the colluvial fan | 320 | | 15.8 |
| material from hilslope erosion deposited on the concave hillslope | 270 | | 14.6 |
| gully erosion | | 930 | |
| total amount | 590 | 930 | 30.4 |

# 8   Discussion

## 8.1   The influence of land use on soil erosion

The human impact on local and regional cultural landscape development and vegetation history has varied greatly in time and space over the last 5000 years (Birks et al., 1988b). Thus the influence of land use on soil erosion analysed at the Biesdorfer Kehlen is partly due to local activities, but it follows regional agricultural trends as well. This was clearly shown for the second half of the 20$^{th}$ century, where industrialized and standardized agriculture facilitated gully erosion.

Historical documents, indicating historical land use in Northeast Germany are rare. No information about the land use history have been found in archives for the investigation site. Sources of land use changes on the regional scale can be found in archaeological and historical literature (eg. Kamke, 1996; Lutze, 1994, for Brandenbug). This information provides a general view of the land use tradition for a certain time but says little about what was practised in the investigated area. However, general information about land use trends and population development can be linked and compared with the stratigraphic results of analysed geoarchives, resulting in a reconstruction of the human impact and its causes.

Analyses of geoarchives are a source for the reconstruction of land use in prehistoric times. Schatz (2000) describes for northeast Brandenburg local and temporary soil erosion on used areas in Neolithic Times and the Bronze Age. From late Bronze Age until the beginning of the Early Middle Ages the increase in population led to an intensification of land use and subsequently to an increase in soil erosion. The local influence of the Bronze Age and Iron Age settlements on the landscape can be illustrated with the following examples: Near the village Rathsdorf, about 5 km north of the nature reserve Biesdorfer Kehlen, soil erosion facillitated by an Iron Age settlement took place (Schmidtchen at al., 2002b). At the Biesdorfer Kehlen the gully and hillslope erosion occurred from the Neolithic until the Bronze Age. From Iron Age until early Medieval Times 3000 years of geomorphic stability and soil formation under woodland vegetation took place. During the Early Middle Ages (300–700 AD) soil formation under woodland vegetation was the main process for the landscape development in wide parts of Brandenburg (Schatz 2000, Schmidtchen et al. 2002a).

Bork and Lang (in this volume) describe the land use and soil erosion development from the Early Middle Ages until today for Germany as follows: In the middle of the 6$^{th}$ century, 90 % of the surface was covered with woodland. In the early 14$^{th}$ century woodland coverage in Germany was reduced to 15 % of the land surface area. Erosion increased drastically. The worst catastrophe took place in 1342. Fields had to be abandoned and resulting starvation and the onset of the black death in 1348/50 killed more than a third of Germany's population. The abandoned areas regenerated with woodland, soil erosion played a minor role in landscape development. In the 16$^{th}$ century, farming and soil erosion increased again. In the late 16$^{th}$ century one third of Germany was covered with woodland which were used intensively for grazing. Soil erosion increased again drastically during the 17th and 18th centuries. In the second half of the 20th century the soil erosion rates again became high caused by changing crop sequences, increasing field sizes and mechanisation of agriculture.

Similar results for Medieval Times and Modern Times have been published by Schatz (2000) and Schmidtchen (2002a) for study areas in northeast Brandenburg. The comparison with the example Biesdorfer Kehlen shows similarities but also some differences from the described pattern. After a long period without human impact –from Iron Age until early Medieval Times – the first trace of woodland clearing was found when the Slavonic people used the area for agriculture again. This could be proved by marks of mouldboard ploughing in a humic–horizon. In contrast to several other sides nearby, for example the Wolfsschlucht about 15 km southwest of the Biesdorfer Kehlen, only few deposits from hillslope erosion could be detected for High and late Medieval Time at the Biesdorfer Kehlen. At the Wolfsschlucht the main gully erosion phases took place in late Medieval Times and in the second half of the 18th century. For example, in the second half of the 14th century one erosion event extended the gully Wolfsschlucht by 42 m (Schmidtchen et al., 2002a). The lack of significant soil erosion from the 15th until the 17th centuries at the Biesdorfer Kehlen can be explained by a dense vegetation cover that prevented the soil from erosion during heavy rainfall.

At the nature reserve Biesdorfer Kehlen intensive land use facilitated severe hillslope erosion and gully erosion during Modern Times until 1991. The drastic gully erosion in the 17th and 18th centuries and in the second half of the 20th century were facilitated by an intensive land use. Compared with the above mentioned results from Bork and Lang (in this volume) and with the results from other investigation areas in Brandenburg (Schatz, 2000; Schmidtchen, 2002a) these were widespread phenomena for these periods of time.

## 8.2   The influence of climate on soil erosion

Land use facilitates soil erosion but extreme events like heavy rainfall, storm or warm air leading to a rapid snow melt, cause soil erosion (Bork et al., 1998).

Therefore, reconstructions of past climate trends are essential for the interpretation of data derived from the analysis of geoarchives. Problematic is the reconstruction of the local climate or even more difficult the reconstruction of heavy rainfall events that led to soil erosion in a certain investigation area. Several methods can be used. For example the reconstruction of rainfall events from contemporary documents. Unfortunately hardly any historical documents are available for the investigation area Biesdorfer Kehlen. Some documents mention floods of the river Oder from 1500 until today (Schulz, 1998). They do not attest heavy rainfall at the lower part of the river, where the investigation area is situated. No other source of historical local climate or precipitation reconstruction can be obtained for the investigation area.

But the dated stratigraphy at the Biesdorfer Kehlen reflect some of the known climate changes from Neolithic until today. The correlation of the results with the climatic reconstruction published by Schönwiese (1995) shows that the first prehistoric erosion phase took place in the cold and wet subatlantic period. There has been little erosion in the late Medieval Times at the Biesdorfer Kehlen, when the climate shifted to a period known for heavy rainfall in Central Europe (Glaser, 2001; Schönwiese, 1995) and intensive soil erosion (Bork et al., 1998; Schatz, 2000).

The second main erosion phase at the investigation area took place 17th and 18th centuries. Namely during the 2nd half of the 18th century heavy rainfall caused dramatic gullying (Hard, 1976).

For the second half of the 20th century a correlation of measured precipitation data, land use and soil erosion is possible. But even the correlation between soil erosion events and rainfall events for the last decades is difficult, because of the spatial and temporal difference in rainfall distribution during this period (Schmidtchen et al., 2001). The correlation of precipitation data and soil erosion events was possible because of the combination of several methods that allowed the exact dating of the sedimentary record.

## 9 Conclusions

Information preserved in the sedimentary records of geoarchives is important to provide information about past climate impact and land use impact on the studied catchment areas. Geoarchives contain information about the land use of a certain time. For the example shown in this contribution dense vegetation prevented dramatic soil erosion during the late Medieval phase known for intensive rainfalls and soil erosion. Geoarchives also contain information about past precipitation events. At the Biesorfer Kehlen intensive precipitation caused severe gully erosion during prehistoric times, the 17th, 18th, 19th, and 20th centuries.

The reconstruction of younger erosion events (20th century) from geoarchives can be used for calibrating and validating models that simulate past erosion and landscape changes. The correlation of measured precipitation data, exact land use data and dated single soil erosion events lead to an exact reconstruction of the interactions of relevant factors facilitating landscape changes in a catchment area.

### Acknowledgement

The following persons are gratefully acknowledged: M. Dotterweich, A. Erber, C. Dalchow, K.–U. Heussner, P. Grootes, A. Manzano, Chr. Gödicke, U.–K.Schkade, Mo. Frielinghaus, Y. Li, and M. Naumann, the many students of the University of Potsdam

## References

AG Boden (1994): Bodenkundliche Kartieranleitung. 4. Edition, Hannover.

Bauer, A. W. (1993): Bodenerosion in den Waldgebieten des östlichen Taunus in historischer und heutiger Zeit – Ausmass, Ursachen und geoökologische Auswirkungen. Frankfurter Geowissenschaftliche Arbeiten, Serie D Physische Geographie, Band 14.

Birks, H. H., Birks, H. J. B., Kaland, P. E., and Moe, D. (1988a): The Cultural Landscape – Past, Present and Future. Cambridge University Press.

Birks, H. J. B., Line, J. M., and Perrson, T. (1988b): Quantitative Estimation of Human Impact on Cultural Landscape Development.–In: Birks, H. H., Birks H. J. B., Kaland, P. E., and Moe, D. (eds.): The Cultural Landscape – Past, Present and Future. Cambridge University Press.

Bork, H.-R., and Dalchow, C. (2000): 4.1. Reliefaufnahme.–In: Barsch, H., Billwitz, K., and Bork, H.-R. (eds.): Arbeitsmethoden in Physiogeographie und Geoökologie. Klett–Perthes, Gotha and Stuttgart, 143–172.

Bork, H.-R., Bork, H., Dalchow, C., Faust, B., Piorr, H.-P., and Schatz, T. (1998): Landschaftsentwicklung in Mitteleuropa: Wirkungen des Menschen auf Landschaften. 1. Ed., Gotha, Klett–Perthes.

Dearing, J. (1991): Erosion and land use. –In: Berglund, P. E. (Ed.): The cultural Landscape during 6000 years in southern Sweden – the Ystad Project, Ecological Bulletin 41: 283–292.

Dincauze, D. F. (2000): Environmental Archaeology – Principles and Practice. Cambridge University Press, Cambridge.

Edwards, K. J., and Whittington, G. (2001): Lake sediments. Erosion and landscape change during the Holocene in Britain and Ireland. Catena 42: 143–173.

Franz, H.-J., Schneider, R., and Scholz, E. (1970): Geomorphologische Übersichtskarte 1:200.000 – Erläuterungen für die Kartenblätter Berlin–Potsdam und Frankfurt–Eberswalde. Gotha/Leibzig, VEB Hermann Haak.

Glaser, R. (2001): Klimageschichte Mitteleuropas – 1000 Jahre Wetter, Klima, Katastrophen. Darmstadt, WBG.

Hard, G. (1976): Exzessive Bodenerosion um 1800.–In: Richter, G. (ed.) Bodenerosion in Mitteleuropa, Wege der Forschung 430: 195–239.

Kamke, H.-U. (1996): Barnim und Lebus. Studien zur Entstehung und Entwicklung agrarischer Strukturen zwischen Havel und Oder, Deutsche Hochschulschriften 1106.

Ludwig, B., Boiffin, J., Chadoeuf, J., and Auzet, A.-V. (1995): Hydrological structure and erosion damage caused by concentrated flow in cultivated catchments. Catena 25: 227–252.

Lutze, G. (1994): Geschichte der Landnutzung.–In: Bork, H.-R., Dalchow, C., and Frielinghaus, Mo. (eds.): Exkursionsführer Nordost–Deutschland und Westpolen. ZALF–Bericht 14, Müncheberg: 86–91.

Marchinek, J. and Nitz, B. (1973): Das Tiefland der Deutschen Demokratischen Republik – Leitlinien seiner Oberflächengestaltung. Gotha/Leibzig, VEB Hermann Haak.

Mölders, N. (2000): Similarity of Microclimate as Simulated in Response to Landscapes of the 1930s and the 1980s. Journal of Hydrometeorology, Volume 1: 330–352.

Munsell (2000): Munsell Soil Color Charts. GretagMacbeth, NY.

Pfister, Ch. (1999): Wetternachhersage: 500 Jahre Klimavariationen und Naturkatastrophen (1486 – 1995). Bern, Verlag Paul Haupt.

Rapp, G. Jr., and Hill, Chr. L. (1998): Geoarchaeology – The Earth–Science Approach to Archaeological Interpretaion. Yale University Press, New Heaven and London.

Rössner, U. and Töpfer, Chr. (1999): Historische Bodenerosion auf Flurwüstungen im westlichen Steigerwald. Mittlg. Fränk. Geogr. Gesellschaft Bd. 46, Erlangen: 27–74.

Schatz, T. (2000): Untersuchungen zur holozänen Landschaftsentwicklung Nordostdeutschlands. Zalf–Bericht Nr. 41, Müncheberg.

Seils, M. (2000): Holozäne Sediment– und Bodenverlagerungen im östlichen Harzvorland. Wirkungen und Ursachen nutzungsbedingter Landschaftsveränderungen, Trift Verlag, Halle.

Semmel, A. (1995): Development of gullies under forest cover in the Taunus and Crystalline Odenwald Mountains, Germany. Z. Geomorph. N.F., Suppl.–Bd. 100: 115–127.

Schmidtchen, G., Bork, H.–R., and Dotterweich, M. (2001): Junge Schluchtenerosion in Ostbrandenburg. Petermanns Geographische Mitteilungen 2001/6: 74–82.

Schmidtchen, G., Dotterweich, M., and Bork, H.–R. (2002 a): Untersuchungen zur Holozänen Genese des Narurparks Märkische Schweiz. Forschungen zur Deutschen Landeskunde (in press).

Schmidtchen, G., Govedarrica, B., and Bork, H.–R. (2002 b): Besiedlung, Bodenerosion und Wasserhaushalt – Die Entwicklung einer Insel im Oderbruch bei Rathsdorf in Ostbrandenburg. Forschungen zur Deutschen Landeskunde (in press).

Schkade, U.–K., Frielinghaus, Mo., Li, Y., and Naumann, M. (1999): Bestimmung des Gehaltes an Cäsium–137 und natürlichen Radionukleiden in Bodenproben aus der Umgebung von Wriezen. Bericht ST2–80/1999 des Bundesamtes für Strahlenschutz, Fachbereich ST, Berlin/München.

Schönwiese, Chr. (1995): Klimaveränderungen – Daten, Analysen, Prognosen. Springer Verlag Berlin, Heidelberg.

Schulz, R. (1998): Betrachtungen zur Besiedlungsgeschichte im mittleren und unteren Oderbruch anlässlich des Oder–Hochwassers 1997. Beiträge zum Oderprojekt 4, Deutsche Archäologisches Institut, Römisch–Germanische Kommission, Berlin: 117–123.

Souchere, V., King, D., Daroussin, J., Papy, F., and Capillion, A. (1998): Effects of tillage on runoff directions: consequences on runoff contributing area within agricultural catchments. Journal of Hydrology 206: 256–267.

Starkel, L. (1987): Anthropogenic Sedimentological Changes in Central Europe. STRIAE 26: 26–29.

Takken, I., Govers, G., Jetten, V., Nachtergaele, J., Steegen, A., and Poesen, J. (2001): Effects of tillage on runoff and erosion patterns. Soil and Tillage Research 61: 55–60.

# Land Use and Soil Erosion in northern Bavaria during the last 5000 Years

Markus Dotterweich

Ökologie–Zentrum der Christian–Albrechts–Universität zu Kiel, Olshausenstrasse 40, 24098 Kiel, Germany

**Abstract.** To reconstruct past environmental conditions and feedback mechanisms between human activities and the environment, geoarchives are an important scientific source of information. Especially colluvial sediments originating from soil erosion can be used to explain environmental effects of land use. Here, results of two case studies from adjacent sites in northern Bavaria are presented. The influence of weather and land use effects on accelerated soil erosion and the feedback on socio–economic and demographic effects are discussed. At "Catena Friesen", an 80 m long slope, the effects of land use changes and their consequences on soil erosion and soil fertility are demonstrated. Several phases of activity and stability from the Latest Neolithic to Modern times were distinguished and soil erosion quantified. The highest amount of soil loss took place in the Medieval Times by two single erosion events. With the second example "Hainbach" the development of a 400 m long and up to 6.4 m deep gully system is described. The main phases of gully erosion took place in the 10th century and between the 14th and 18th centuries. During this time, 15 % of the catchment area was gullied. At both locations, intensive land use led to soil erosion. The extreme soil erosion in the medieval period, however, was caused by extraordinary weather phenomena. Additionally, land use patterns were responsible for the development of a gully system at Hainbach.

## 1  Introduction

The extent of human impact on the landscape has changed considerably since the first farmers cleared woodland during the Neolithic period in order to practice agriculture. Phases of expansion and regression or clearance and reforestation occurred. They included changes in human settlement and land use patterns, population density, climate, vegetation composition and structure, local or regional hydrology, and soil fertility, as well as socio–economic, demographic, and cultural factors, and the complex but poorly understood interactions between these components (cf. Birks et al. 1988, Bork et al. 1998).

Information on landscape development is partly stored in geoarchives. Among others, these are colluvial fans, deposits on footslopes or gully in–fills in which eroded soil material has been deposited. The clearing of forest and subsequent agricultural land use provided appropriate conditions for soil erosion. Heavy rainfall led to high runoff on the surface and soil erosion occurred. Under a forest cover, in Central Europe the annual soil loss appears to be small and insignificant. However, today thousands of gullied slopes and colluvial fans can be found in the landscape, mainly under forest. They indicate that past soil loss had dramatic effects on topography and soil fertility. Today, the long–term effects of soil erosion are largely unknown. Also, little is known about the intensity and frequency of past soil erosion events.

To reconstruct and understand the interactions between historical land use and soil erosion detailed investigations on the formation, volume and age of sediments and the soil development in geoarchives (res. colluvial deposits) are essential. Similarly, information about the history of the local land use patterns and local or regional extreme precipitation events are necessary. To obtain high–resolution quantitative data on historical soil loss studies in small catchments are required (Bork et al. 1998, Dotterweich 2002). Such information can thus be used to calibrate numerical models for historical soil erosion, landscape evolution, and nutrient and energy fluxes over longer periods of time. This should provide new insights into the processes of landscape development.

Only few field studies on land use effects to the landscape development in northern Bavaria exist: Becker (1983), Schirmer (1983, 1988) and Gerlach (1990) illustrate the response of the rivers Main and Regnitz on climate and land use dynamics. They show increased reworking of Holocene terrace sediments since Medieval times coinciding with the maximum extent of the Alpine glaciers during the Little Ice Age. It is assumed that for the river dynamics land use plays a modifying role only.

Field studies of smaller catchments investigating land use and soil erosion in Medieval and Modern times in northern Bavaria were carried out by Machann et al. (1970), Abraham de Vazquez et al. (1985), Hildebrandt et al. (1993), Garleff (1988, 1989), Rösner et al. (1999), Schmitt et al. (2002) and Dotterweich et al. (2002b, 2002c). The locations Rambach, Schmerb, Obersteinach, Dürrnitz and Winkelhof are abandoned fields in the Steigerwald area, around 30 km southwest of Bamberg (Table 1). The comparison of soil thickness at truncated soil profiles with pristine soil profiles nearby indicated a soil loss of up to 60 cm. At the Ellernbach valley some kilometers east of Bamberg a soil loss of 1 m in around 1000 years could be determined from sediments in the valley floor. At the Wolfsgraben, it was possible to reconstruct and quantify soil erosion and sediment mobilisation for different phases: The highest soil erosion phases took place in the Late Medieval Times and in the $18^{th}$th/$19^{th}$ centuries. This affected gullying in the Wolfsgraben. In between and in the $20^{th}$ century soil erosion caused soil losses of about 7 cm on the slopes.

Unfortunately, the results are hard to compare because different methods and scales were used. The crucial point of discussion is about the dominant factor causing accelerated soil erosion in the $14^{th}$ and $18^{th}$ centuries. Are heavy rainfall events, i.e. climate factors, or land use effects responsible?

In this contribution, results from two local studies from the Triassic hill country in northern Bavaria are presented. In the first example, Catena Friesen,

**Table 1.** Long term soil erosion rates for different sites in northern Bavaria.

| Location | Coordinates | Period | Slope length (m) | Slope angle | Soil loss (cm) | Reference |
|---|---|---|---|---|---|---|
| Rambach | 42°52'/10°35' | ~850 – 1325 AD | 1050 | 1–6.5° | 8–28 | Rösner et al. 1999 |
| Schmerb | 42°52'/10°31' | ~1317 – 1868 AD | 570 | 0.5–8° | 23 | Rösner et al. 1999 |
| Obersteinach | 42°50'/10°34' | ~1281 – 1909 AD | 360 | 1.5–6.5° | 17 | Rösner et al. 1999 |
| Dürrnitz | 42°45'/10°23' | ~1300 – Modern Times? | no data | no data | 10–30 (–100) | Machann et al. 1970 |
| Winkelhof | 42°49'/10°31' | 1325 – ~1400 | no data | 1–7° | 18 | Hildebrandt et al. 1993 |
| Ellernbach valley | 42°55'/11°02' | ~150 BC – ~400 AD and ~1400 AD – ~1900 AD | ~500 – ~1000 | 1–15° | 100 | Abraham de Vazques et al. 1985, Garleff 1988, 1989 |
| Wolfsgraben | 49°57'/10°52' | ~1350 – ~1700 | 100 | 10–20° | 7* | Schmitt et al. 2002, Dotterweich et al. 2002b,c |

* Excluding hillslope erosion during the gullying phases in the first half of the 14th and in the 18th century.

slope deposits are analysed to determine changes in landform caused by soil erosion during the last 5000 years. The site was chosen due to the opportunities offered by an archaeological excavation of an abandoned medieval settlement located nearby. The second example Hainbach shows the development of a gully system 400 m in length and up to 6.4 m in depth by analysing the gully infill.

## 2   Materials and Methods

To adequately analyse temporal and spatial changes of a landscape, several methods from different disciplines are required. The Task Force of the European Society for Soil Conservation (ESSC) on long–term effects of land use on soil erosion in a historical perspective developed a methodological procedure for the reconstruction of past soil erosion events and their causes, which was applied here.

### 2.1   Study area:

The studied sites are situated in the hill country between the Obermaintal (upper valley of the Main river) in the south and the Frankenwald region in the north at about 320 to 450 m a.s.l. (Figure 1). The study area has a complex structure of different fracture blocks composed by lithified limnic–fluvial sandy and clayey layers from the Triassic period (Dannapfel 1991, Emmert et al. 1972, Geologische Karte von Bayern 1972, sheet 5734). The valleys of the rivers Kronach, Haßlach and Rodach draining into the river Main cut these layers. Sandy–silty periglacial solifluction layers cover most of the slopes. Quaternary mass movements often cover the footslopes and thick stacks of fluvial deposits are being found in the valleys (Emmert et al., 1972). Today, the climate is humid and the average annual precipitation ranges from 750 to 800 mm and the mean annual temperature ranges from 7 to 8° C (Klimaatlas von Bayern 1996).

The climate history of Central Europe including northern Bavaria shows variations during the last 5000 years (Schönwiese 1995): After the warm and wet climate dominating in the Atlantic period a cool and dry phase followed in the Subboreal period. Precipitation increased at the beginning of the cooler Subatlantic period. The period of the Roman Empire was characterised by a warm and wet climate. A further cool and wet period in the Dark Ages followed. In the Early and High Medieval period temperatures increased again and it was temporary dry. In the 13th and 14th century cool and warm years alternated and heavy rainfalls occurred. This is also shown by many historical documents describing huge floods (especially in the first half of the 14th century) caused by widespread heavy rainfall events in northern Bavaria (Glaser 1991, 2001). The largest rainfall event in the last millennium occurred in July 1342 (Bork et al. 1998, Glaser 2001, p. 66). From the 15th century until the second half of the 19th century, temperatures decreased. During this period intensive precipitation events occurred. Especially in the 18th century many extreme rainfall events and large floods took place in Central Europe (Bork et al. 1998). Contemporary

**Fig. 1.** Research area

documents describe large floods and intensive rainfalls in northern Bavaria in the years 1431, 1433, 1451, 1551, 1552, 1613, 1718, 1719, 1720 and 1732. (Weikinn 1958–1963 and Glaser 1991, 2001). The earliest report of an extreme flood of the river Kronach dates back to the year 1552: "this wild and large flood caused damage, tore down and to some extent totally destroyed 40 houses, barns and bridges, foot bridges and other communal stone and timber buildings on the Haßlach and Kronach, not to mention the great damage to mills, pastures, fields, and timber, as well as boards, soils and pathways (...)" (Stöhr 1825, p. 245).

Arable land and grassland dominate the current land use on the gentle slopes and partly also the hill tops. Most of the ridges and the steeper slopes are covered by forest. The valley bottoms are used widely as grassland and fields.

In northern Bavaria, agricultural land use started in the Neolithic period ca. 5000 years ago (Züchner 1996). In the region of Kronach evidence of settlements from the Bronce Age until the Roman Times were found. In the Dark Age the population and the agriculturally used areas decreased strongly in northern Bavaria (Sage 1996). Heinold–Fichtner (1951) and Demattio (1998) describe the settlement history in the investigation area since the Medieval period: In the 9th and 10th centuries, the first medieval villages were founded in the region of Kronach. In the 12th and 13th centuries, many parts of the forests were cleared and numerous villages were founded. In the first half of the 14th century, many villages in northern Bavaria were abandoned including nearly all the villages around the investigated sites (Friesen, Glosberg, Vonz, Birkach and Dörfles (Figure 1). v. Geldern–Crispendorf (1930: 89) notes that up to 53 % of the villages in the Frankenwald region were abandoned during the 14th and the middle of the 15th century. Most of these villages were resettled by the end of the 15th century. A second decrease in population occurred during the Thirty Years War between 1618 and 1648 (Endres 1990). In the 18th century, the population and the intensity of land use increased again. The oldest large scale maps from the study area are from the 19th century. They show that at the study sites land use was dominated by grassland at that time (Catena Friesen: Urmesstischblatt NW 102–07 from 1853; Catena Hainbach: Urmesstischblatt NW 102–08 from 1853, Landesvermessungsamt Bayern). In the second half of the 20th century, the intensity of agricultural land use further decreased. Many fields changed into grassland or forest also at Catena Friesen: In 1961, most of the fields were transformed into permanent pasture and the slope opposite of Catena Friesen changed from arable land into forest. In 1991, a second reforestation took place on the lower part of the investigated slope (pers. comm. H. Wich, Friesen, 03.10.2000).

## 2.2  Sediment analyses:

To investigate sediment stratification and soil development it was necessary to excavate large trenches to obtain maximum information from the geoarchive. After cleaning the sides of the trenches, the limits of sediment layers and soil horizons were identified. To determine further details on soil development, soil samples were taken for field analysis of texture, pH value, and colour (AG Boden 1994, Munsell 1975). Carbonate content and organic matter content were analysed in the laboratory (AG Boden 1994).

## 2.3  Dating:

Sediment samples and remains of potsherds, wood and charcoal were collected for dating. Large and un–rounded pieces of charcoal and potsherds were selected to avoid sampling of reworked material. The youngest age obtained on the embedded material was taken as closest representation of age of the sediment.

Radiocarbon dating of wood and charcoal were carried out by P.M. Grootes and H. Erlenkeuser at the Leibniz–Laboratory, University of Kiel, Germany. In this paper only calibrated ages are used (at the $2\sigma$ confidence level). Ages of pottery fragments were obtained using two methods: Archaeological classification carried out by J. Haberstroh and H. Endres from the Bavarian State Office for Heritage Protection. Archaeological nomenclature was taken from Sage (1998). Thermoluminescence (TL) dating carried out by Ana Manzano at the Rathgen–Forschungslabor, Berlin, Germany. One sediment sample at Catena Friesen was dated by optically stimulated luminescence (OSL) by B. Mauz at the Geographisches Institut, University of Bonn.

## 2.4  Quantifying soil loss:

Average rates of soil erosion were calculated for various phases by

1. determining the accumulation slope lengths of each soil erosion phase ($C_L$ [m]);
2. calculating the mean thickness of stored sediment on the slope ($C_M$ [m]);
3. calculating the cross section area of sediment accumulated on the slope ($C_A = C_L C_M$ [m²]);
4. determining the erosion slope length assuming a constant runoff contributing area of each soil erosion phase ($E_L$ [m]);
5. calculating the average lowering of the eroded slope of each soil erosion phase ($E_M = C_A / E_L$ [m]);
6. extrapolating the average lowering to a hypothetic erosion surface of 1 ha;
7. applying a density $\rho$ [1.6 g cm$^{-3}$] for the former soil and
8. quantifying the annual soil erosion rates by $A_a = E_M 10000 \rho / a_a$ [t ha$^{-1}$ a$^{-1}$]).

Years ($a_a$) are maximum number of years with possibility of soil erosion derived from the stratigraphy and the land use period. Because of the knowledge gaps in the land use history, the numbers of these years are estimates, in particular for pre–historical periods. Phases with single soil erosion events were calculated as one particular year. With such an approach, it has to be taken into account that the derived erosion values represent spatial mean values only. It is assumed that highest soil removal occurred on the steepest, most convex parts of the slope and the smallest soil removal on the crest. It should also be taken into account that parts of the eroded soil were transported out of the slope system. Therefore, the calculated values represent minimum values only. It is assumed that the actual soil erosion was 10 to 30 % higher.

# 3   Study sites

**Catena Friesen:**

A few kilometres north of the town of Kronach, approximately 1 km west of the village Friesen, Middle Age potsherds were discovered on farmland in 1990 (50°17 Latitude, 11°22 Longitude ca. 370 m above seal level, see Figure 1). This initiated an archaeological excavation by the Bavarian Office for Heritage Protection in 1992. This excavation, which continued until 2000, extended over approx. 2 ha on the crest and upper slope of a small hill. South of the archaeological excavation, a slope extends over approx. 80 m with a height difference of about 7 m. Today it drains into a small ditch that conveys the water into fishponds near the village Friesen. At present most of the slope is under pasture, only the lowest part is covered with pine forest. Along the slope, six soil pits were excavated, with a total length of 55 m (Figure 2).

Few evidences of land use change during the Neolithic, the Bronze Age period and three settlement phases with local field abandonment in the Middle Ages were revealed by archaeological investigations. Remains of poles from ground level timber constructions and pit houses (Grubenhäuser) were dated to the first medieval phase between 800 and 1000 AD. The archaeological excavations

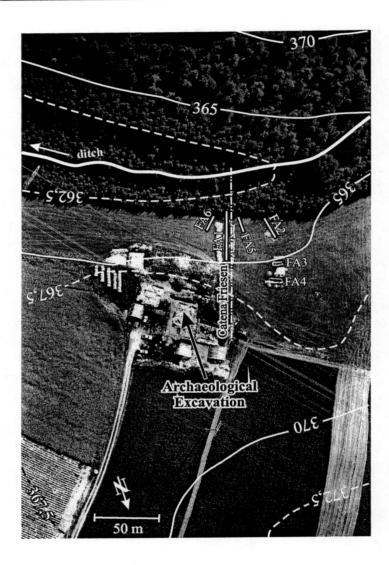

**Fig. 2.** Location of Catena Friesen (Aerial photograph taken at 24.08.2001 with friendly permission from the Bayerisches Landesamt für Denkmalpflege, Abteilung Luftbildarchäologie, picture no. 5734/011-1:7857-19) contours: m.a.s.l.; FA1 - FA6: exposures.

revealed two square stone towers, each 10.5 x 8.5 m in size, from the High Medieval period at the upper slope and on the hilltop. According to potsherd finds, the breakdown of the towers must have taken place at the latest at around 1200 AD. Ground level timber constructions with a cross section of up to 1.2 m mark the third and final Middle Age settling period in the middle of the 14th century (Dotterweich et al. 2002a).

### The gully system at Hainbach:

In the hill country developed in Triassic sediments north of the city Kronach, at least 10 gullies have cut into the western and eastern slopes of the river valleys Kronach and Haßlach. One of them, the gully system Hainbach, around 2 km north of Kronach, was investigated (Figure 1).

Some houses in the western part of the village Dörfles, approx. 3 km north east of Kronach, are constructed on a clearly visible fan of approximately 2–3 m in height. Here the Hainbach valley ends (50°16 Latitude, 11°20 Longitude, approx. 360 m above sea level, see Figure 1). The valley is around 1000 m long and a small creek flows in it and drains into the river Kronach. Today, the steep southern slopes are predominantly used as pine forest, whereas almost all northern and western slopes are used as permanent pasture.

The upper part of the steep western slope is covered by a copse (ca. 1 ha). Within the forest, two gullies run downslope with a depth of up to 3.5 m and a width of 20 m (gullies 2 and 3, see Figure 8 today). The gullies cut into a solifluction layer and the underlying weathered sandy–clayey Triassic bedrock. On the same slope, but outside the forest, a depression running downslope is visible (gully 1, Figure 8). It is 1 to 1.5 m deep, up to 15 m wide and has a length of around 100 m. Close to the valley floor this depression joins the common gutter of gullies 2 and 3.

## 4   Results

### Catena Friesen

Seventeen stratigraphic layers were differentiated and are shown in Figures 3 and 4. The characteristics of the sub–soil (layers 1 and 2), the colluvial deposits (layers 3 to 12 and 14 to 16) and the in–fill of pits (layers 13 and 17) are described in Table 2 and below. The dating results obtained on charcoal, potsherds and sediment are shown in Table 2 and Figure 5.

In the upslope area, directly below the ploughing horizon, weathered Triassic bedrock (layer 1) is found in some places. Further downslope, the Triassic bedrock is covered with a sandy–silty solifluction layer up to 1 m in thickness (layer 2). Some in–filled ice wedges are visible within layer 2. A few centimetres beneath the upper border of this layer a piece of charcoal was found and dated to BC 7296 - BC 7051 (Table 2).

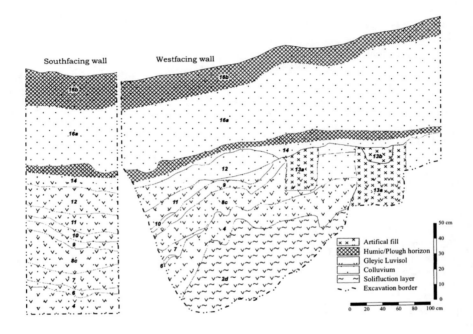

**Fig. 3.** Catena Friesen: Sketch of exposure FA6.

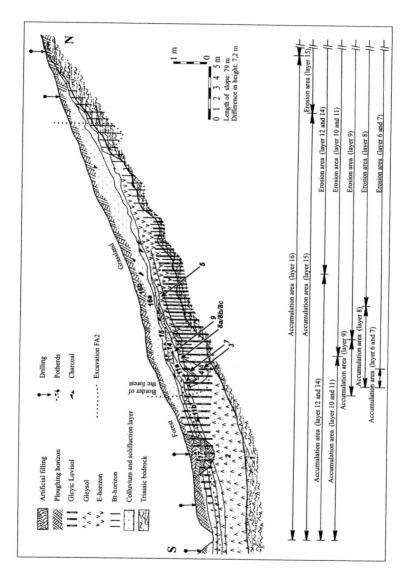

**Fig. 4.** Sediments and soils at Catena Friesen. The location of exposure FA 2 is indicated, as are the locations of drillings. Numbers plotted correspond to layer numbers in the text. Uphill the slope continues another 33 m until the crest. In the lower part, accumulation and erosion areas are indicated for different sediment layers.

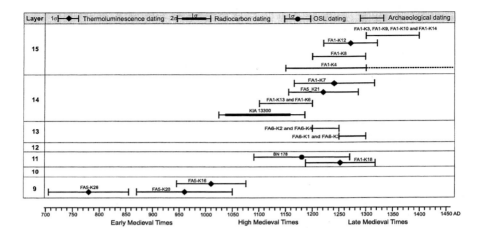

**Fig. 5.** Ages of charcoal and potsherds from layers 9 to 15 at Catena Friesen.

Table 2: Sediments, soils and age from Catena Friesen.

| No. | Description[1] | Colour[2] | Texture[3] | Sample No., material used | Age[4] |
|-----|----------------|-----------|------------|---------------------------|--------|
| 1 | Bedrock, Lower Trias (Kulmbacher Konglomerat); in places: banded clay illuviation, gleyic luvisol, and gley | 7.5 YR 6/8; 10 R 3/2 | sand | | |
| 2 | Solifluction layer, in-filled ice wedges, clay illuviation; in places: banded clay illuviation, gley or gleyic luvisol | 2.5 Y 6/4; 10 YR 6/4 | silty sand | KIA 14136 (charcoal) | AMS: BC 7296 – BC 7051 (BP 8112 ±40; $\delta$13C: -23.20) |
| 3 | Colluvium, gleyic luvisol, many pieces of charcoal | 2.5 Y 5/1 | sand | KIA 11582 (charcoal) | AMS: BC 2011 – BC 1742 (BP 3539 ±42; $\delta$13C: -24.37) |
| 4 | Colluvium, gleyic luvisol, few pieces of charcoal | 2.5 Y 5/1 | sand | | |

Table 2: Sediments, soils and age from Catena Friesen.

| No. | Description[1] | Colour[2] | Texture[3] | Sample No., material used | Age[4] |
|---|---|---|---|---|---|
| 5 | Colluvium, gleyic luvisol, in-filled ice wedges | 2.5 Y 6/4 | silty sand | FA5-K27 (potsherd) | AA: late Bronze Age |
| 6 | Colluvium, alternating clay and sand layers, gleyic luvisol | 10 YR 6/2 | clay and sand | | |
| 7 | Colluvium, luvisol, many pieces of charcoal up to 2 cm in diameter | 10 YR 4/2 | loam | KIA 12312 (charcoal) | AMS: BC 793 – BC 415 (BP 2510 ±30; $\delta$13C: -30.97) |
| | | | | KIA 14131 (charcoal) | AMS: BC 785 – BC 411 (BP 2484 ±30; $\delta$13C: -25.87) |
| | | | | KI 4900 (charcoal) | Beta: BC 765 – BC 405 (BP 2470 ±35; $\delta$13C: -26.33) |
| 8a | Colluvium, clay eluviation | 10 YR 7/2 | silty sand | | |
| 8b | Colluvium, clay eluviation, gleyic luvisol | 10 YR 7/2; 10 YR 4/2 | silty sand | | |
| 8c | Colluvium, gleyic luvisol, tree stumps | 10 YR 7/2 | silty sand | KI 4876 (tree stump) | Beta: AD 85 – AD 325 (BP 1830 ±25; $\delta$13C: -26.14) |
| | | | | BN 178 (sediment) | OSL: 1800 ±200 a |
| 9 | Colluvium, gleyic luvisol, charcoal | 2.5 Y 5/4 | loam | FA5-K20 (potsherd) | TL: 870 – 1050 AD |
| | | | | FA5-K26 (potsherd) | TL: 685 – 855 AD |
| | | | | FA1-K16 (potsherd) | TL: 945 – 1075 AD |
| 10 | Humic colluvium, many pieces of charcoal | 7.5 YR 3/1 | sand | | |
| 11a | Colluvium derived from a luvisol, some pieces of charcoal | 7.5 YR 4/1 | sand | FA1-K18 (potsherd) | TL: 1181–1321 AD |

Table 2: Sediments, soils and age from Catena Friesen.

| No. | Description[1] | Colour[2] | Texture[3] | Sample No., material used | Age[4] |
|-----|----------------|-----------|------------|---------------------------|--------|
| 11b | Colluvium derived from a luvisol, some pieces of charcoal, gleyic luvisol | 2.5 YR 5/2 ; 7.5 YR 4/1 | sand | | |
| 12 | Colluvium derived from a luvisol, gleyic luvisol | 7.5 YR 4/1 | sand | | |
| 13a | Soil pit in-fill, many pieces of pottery and charcoal | 7.5 YR 2.5/1 | sand | FA6-K2 (potsherd) | AA: 1200–1250 AD |
| | | | | FA6-K3 (potsherd) | AA: 1250–1300 AD |
| | | | | FA6-K4 (potsherd) | AA: 1200–1250 AD |
| 13b | Soil pit in-fill, many pieces of pottery and charcoal | 7.5 YR 2.5/1 | sand | FA6-K1 (potsherd) | AA: 1250–1350 AD |
| 14 | Colluvium derived from a luvisol; pieces of charcoal, humic horizon | 7.5 YR 2.5/1 | sand | KIA 13300 | AMS: AD 1024 – AD 1187 (BP 1020 ±25; $\delta$13C. -27.59) |
| | | | | FA1-K13 (potsherd) | AA: 1000–1300 AD |
| | | | | FA1-K6 (potsherd) | AA: 1000–1300 AD |
| | | | | FA1-K7 (potsherd) | TL: 1170–1320 AD |
| | | | | FA5-K21 (potsherd) | TL: 1150–1290 AD |
| 15 | Colluvium derived from a luvisol, humic horizon | 7.5 YR 2.5/3 | sand | FA1-K3 (potsherd) | AA: 1300–1400 AD |
| | | | | FA1-K4 (potsherd) | AA: 1150–1450 AD |
| | | | | FA1-K8 (potsherd) | AA: 1150–1300 AD |
| | | | | FA1-K9 (potsherd) | AA: 1200–1400 AD |
| | | | | FA1-K10 (potsherd) | AA: 1300–1400 AD |

Table 2: Sediments, soils and age from Catena Friesen.

| No. | Description[1] | Colour[2] | Texture[3] | Sample No., material used | Age[4] |
|---|---|---|---|---|---|
| | | | | FA1-K12 (potsherd) | TL: 1220–1320 AD |
| | | | | FA1-K14 (potsherd) | AA: 1300–1400 AD |
| 16a | Homogeneous colluvium, | 10 YR 3/2 | sand | FA1-K2 (potsherd) | AA: 1700–1900 AD |
| | ploughed | | | FA1-K5 (potsherd) | AA: 1550–1650 AD |
| 16b | Homogeneous colluvium, recent ploughing layer | 10 YR 3/2 | sand | | |
| 17 | recent in-fill (former creek) | – | – | | |

[1] *Geological term from: Emmert et al. (1972); Soil classification after AG Boden (1994).*
[2] *According to Munsell (1975).*
[3] *Determined according to AG Boden (1994).*
[4] **AA:** *Archaeological classification of potsherds.*
**AMS:** *calibrated AMS $^{14}$C-age, $2\sigma$ confidence level (uncalibrated ages in brackets, calibrated using Stuiver et al. (1998)).*
**Beta:** *calibrated conventional $^{14}$C-age, $2\sigma$ confidence level (uncalibrated ages in brackets, calibrated using Stuiver et al. (1998)).*
**TL:** *Thermoluminecence ages according to Aitken (1985), $1\sigma$ confidence level.*
**OSL:** *Optically Stimulated Luminescence age, $1\sigma$ confidence level (values taken from Mauz, et al. (2002)).*

Close to the present day forest border, the in–fill of a creek (2 m wide and 70 cm deep) was found cutting into layer 2. The creek runs approximately parallel to the present forest boarder in east–west direction. The in–fill is a sandy–loamy colluvial sediment (layers 3, 4, 6, 7 and 8). Charcoal was found in layer 3 at the bottom of the in–fill and was dated to BC 2011 - BC 1742 (Table 2). Layer 5, with a thickness up to 10 cm overlays layer 2. Texture and colour of layer 5 are similar to layer 2 with no traces of ice wedge casts. Some potsherds from the late Bronze Age were found in this sediment. The lower limit of this layer is clearly visible only in exposure FA2. Therefore, in Figure 4 this border is symbolised by a dashed line.

The ages of three pieces of charcoal taken from layer 7 are approximately 500 years younger. Remnants of some tree stumps were found standing on layer 7. The youngest tree rings were very poorly developed. The tree stumps are covered with colluvium (layer 8). The radiocarbon age of one tree stump is AD 85 - AD 325 (Table 2).

Colour and structure of layer 9 is typical for a colluvium derived from an E–horizon of a luvisol. Some pieces of potsherds from Early Medieval times were found in this layer (Table 2). On the top of layer 9, a thin and clear distinguishable layer of charcoal and parts of a humic sediment were found (layer 10).

Layers 11 and 12 are colluvial deposits derived from E–, Bt– and C–horizons. Parts of the layers contain some charcoal. A piece of potsherd from the High Medieval period was found in layer 11. The deposition of layer 11 was dated to ca. 1180 AD by OSL (Table 2).

In all excavations, filled fire pits were found, similar to those shown in Figure 3 (layer 13). The pits are full of charcoal and potsherds from Late Medieval times (Table 2). The upper parts of the pit–fills are cut by erosion. This can clearly be seen by a very sharp upper boarder of the pit–fills that represents an erosion surface. The eroded material accumulated on the slope as a black colluvium (layer 14) mixed with material eroded further upslope. This accumulation covers an area of around $400\,m^2$ and has a thickness up to 20 cm.

Layer 15, with a thickness of up to 20 cm extends from the middle to the lower slope. Embedded are many pieces of potsherds from Late Medieval times (Table 2). The uppermost colluvium (layer 16), with a thickness of up to 60 cm was homogenised by ploughing. The age of the embedded potsherds ranges from Medieval times to Modern times. Layer 17 represents an anthropogenic filling with pieces of bricks and sand. It marks the course of a former creek.

In the sediment layers soils developed: In the Triassic bedrock (layer 1) and the solifluction layer (layer 2) at the upper slope, the lower parts of a Bt–horizon of a luvisol were found. Relicts of the luvisols E–horizon with a thickness of up to 0.5 m are preserved in the middle slope area in layers 2 and 8. Also in the archaeological excavation at the top of the hill a fossil luvisol was found covering a pit from the period of the Latest Neolithic/Early Bronze Age (around 2000–1800 BC). This formation lies underneath a medieval cultural layer (cf. Dotterweich et al. 2002a).

In the lower parts of the slope, characteristics of the Triassic bedrock, the solifluction layer and some colluvial deposits as well as parts of the relict luvisol are overprinted by development of gleyic luvisols and gleysols. In the sediments of layers 14 and 15, two A–horizons with thicknesses up to 5 cm developed. Upslope, these horizons are thinning out and finally disappearing.

In layer 16, features of recent ploughing can be found between 0–20 cm depths. Today, at the lowest part of the slope ground water can be found between 50–80 cm below surface.

## The gully system at Hainbach

The gully system at Hainbach is partly filled with sediments. In every gully, exposures were opened across the gully–fills and analysed. In the whole gully

system, more than 150 different sediment layers with many erosion discontinu-
ities were distinguished and a stratigraphic order was established. Details are
presented in Dotterweich (2002).

Figure 6 shows a cross section through gully 1. The location of the cross
section is indicated in Figure 8. The sedimentary structures and the succession
of the layers are typical for all fills studied in the gully system: At the gully
bottom, stony and coarse material dominates. Above that, many different loamy
layers, some with graded bedding, are found. At the top of the filling, brownish
and humic layers composed of clay and silty clay are found. They are compact
and show fine laminations in some places.

Charcoal sampled from the bottom of gully 1 dates from Early Medieval
times. $^{14}$C–ages obtained on charcoal sampled from layers further above are
progressively younger (Figure 6). Most of the fragile charcoal pieces have a large
size and were found in concentrated packages typical for in situ embedded material. Therefore, it is assumed that the ages of the charcoal reflects approximately
the age of deposition of the sediment.

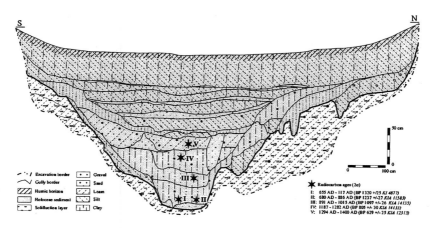

**Fig. 6.** Exposure HA1 in gully 1 at Hainbach. The location is shown in Figure 8.

In contrast to gully 1, charcoal found deepest in excavations at gullies 2 and
3 date to Late Medieval times (Dotterweich 2002). Charcoal pieces sampled from
further above returned ages that are progressively younger. The youngest age
(modern) is obtained on charcoal sampled from near the present surface.

# 5   Discussion

## 5.1   Catena Friesen  Landscape evolution since the late Neolithic period

The following phases of landscape development were reconstructed (Figure 4).

**Phase 1:** Early to middle Holocene: In this period, the climate conditions enabled soil development under natural vegetation on a stable surface. No remnants of this soil can be found today because soil characteristics were

**Phase 6:** Period of the Roman Empire: The woodland on the upper slope was cleared and the geomorphic stability ceased again. Some buried tree stumps on the lower slope, standing on layer 7, indicate this. The trees growing on the lower slope suddenly were partly covered by sediments. For one of the tree stumps an age between 85 and 325 AD was obtained. The sediment layer with a thickness of up to 50 cm indicates intense erosion during this time. Following this event, the partly buried trees began to die; which is indicated by the very poor growths of the youngest tree rings. The good preservation of those tree remnants, suggests that until today they were continuously embedded in an anaerobic environment. No archaeological traces from the Roman Empire were found at the study site. It is assumed that only a short clearing phase on the upslope area and a high runoff event following an intense rainfall caused soil erosion in this time.

**Phase 7:** Dark Ages: In this period, soil formation under natural vegetation on a stable surface was the dominant process. This is indicated by the bleached horizon in the colluvium from the period of the Roman Empire (layer 8). Today, relics of a luvisol are present on the middle slope. On the hill an up to 0.5 m thick E–horizon is preserved. It is assumed that between the Neolithic period and the end of the Dark Ages (around 800 AD) soil development in the upper parts of the Triassic bedrock (layer 1), the solifluction layer (layer 2) and the colluvial deposits (layers 3 to 8) was only interrupted in phases 4 and 6. It is also assumed that clay eluviation and thus the transformation into the luvisol mainly took place in the period between the Roman Empire and the end of the Dark Ages.

**Phase 8:** Accelerated soil erosion in the medieval period:

> **Phase 8.1:** During the first medieval settlement phase woodland was cleared again and soil erosion occurred. This is indicated by 10 cm thick sediment on the lower slope (layer 9) in which potsherds from the period between 800 and 1000 AD are found.

> **Phase 8.2:** Soil erosion during the end of the second medieval settlement phase led to the deposition of a loamy layer, very rich in charcoal (layer 10). The charcoal richness indicates an intensive fire in the catchment before the erosion. It may be closely related to a fire in the settlement and the following breakdown of the medieval stone towers (Dotterweich et al. 2002a). The charcoal in layer 10 was protected from weathering and disturbance by the deposition of further sediment, in which only few potsherds were found. It is assumed that after the breakdown of the stone towers a period of more intense land use followed, soil erosion increased, and layer 10 deposited. This took place almost immediately after the second medieval settlement period (around 1200 AD). Following, land use intensity was still on a high level. Soil erosion led to the deposition of layers 11 and 12. This is also supported by an OSL–dating of layer 11.

> **Phase 8.3:** A single soil erosion event in the third medieval settlement period (13th century and beginning of the 14th century) took place. On the lower slope, a black coloured colluvium rich in charcoal was found (layer 14). The structure of the sediment shows, that the substrate was

accumulated during a single erosion event. Since deposition, a slight humic horizon developed in the sediment. The charcoal found in this colluvium originates from the High Medieval period as most of the potsherds (Figure 5 and Table 2). However, two potsherds are clearly younger and originate from the third medieval settlement period during the 13th century and the beginning of the 14th century. Similar potsherds were also found in the infill of the fire pits (layer 13a and 13b) directly below layer 14. It is assumed that the charcoal and the potsherds found in layer 14 were eroded from the fire pits and transported further down slope during an intensive erosion event in Late Medieval Times. The different ages of the material show that the older potsherds in the colluvium were relocated from the upper slope. The charcoal in the fire pits and in the colluvium have probably the same ages, because they have the same origin. This would show the use of firewood with an age of around 100 years. After this erosion event, land use intensity decreased and the development of a slight humic soil horizon took place, probably under grass cover.

**Phase 8.4:** In the following years, field use on the upper slope was continuing and a second single soil erosion event in the 14th century occurred. A colluvium deposited on the middle and lower slope with a thickness up to 20 cm (layer 15). The sediment structure gives evidence of an accumulation from a single erosion event. The youngest potsherds found in layer 15 originate from the third medieval settlement period. It is assumed that intensive land use and one of the extreme rainfall events in the first half of the 14th century led to high runoff and to soil erosion. Further evidence is provided from historical sources that clearly document many abandoned villages and a dramatic decrease in land use intensity in the region of Kronach during the first half of the 14th century (Dotterweich et al. 2002a).

**Phase 9:** Soil development and soil erosion phases in Modern Times: Contemporary documents indicate that in the surroundings the 15th century farming occurred again. On the site however, the development of humic horizon indicates a stable surface most likely under forest cover. It is assumed that until the 17th century land use in the investigation area was on a low level. Probably, the catchment area was under permanent pasture. A new humic horizon developed on the slope. Then, land use intensified again until 1961. Pasture was transformed into arable land and soil erosion occurred. Accordingly, it is assumed that most of layer 16 was deposited during the 19th and the first half of the 20th century. Many potsherds from medieval times and modern times were found in the sediment. Unfortunately, ploughing homogenised and translocated the sediment and made it impossible to separate further layers.

After the deposition of layer 16 the last landscape change took place at Catena Friesen. Remnants of a former creek on the lower slope (layer 17) were diverted into an artificial ditch to improve the water supply for the fishponds near the village Friesen in the second half of the 20$^{th}$ century.

**Quantification of soil erosion and accumulation at Catena Friesen:** The volumes of sediments in layers 6 to 12 and 14 to 16 were calculated and are listed in Table 3. Since the beginning of agricultural land use in the Late Neolithic until today approx. 70 cm of colluvial sediments accumulated on the middle and lower slope over a distance of 42 m. The upper and middle slope has lost an average of 80 cm of soil over a length of 37 m in this time (Figure 7). Until the end of the Dark Ages, soil formation was the dominant process, interrupted by weak soil erosion phases in the Early Bronze Age, Late Bronze Age, Hallstatt period and the period of the Roman Empire. During the Medieval period soil erosion increased dramatically. In the 14$^{th}$ century two single soil erosion events with a mean slope lowering of 56 and 121 mm ($900\,\mathrm{t\,ha^{-1}\,a^{-1}}$ and $1925\,\mathrm{t\,ha^{-1}\,a^{-1}}$) took place. This represents 74% of the total medieval soil loss and 29% of the total medieval and modern time soil loss. Similar values were derived for the Untereichsfeld region (Bork 1988, p. 88). In Modern Times, the upper slope was lowered by 38 cm. It is assumed that 10% of the sediment was translocated by tillage and 90% by soil erosion (Dotterweich et al. 2002a).

**Table 3.** Soil erosion at Catena Friesen: Number of years with soil erosion ($a_a$), slope length of erosion areas ($E_L$), slope length of accumulation areas ($C_L$), mean thickness of colluvium ($C_M$), cross section of the colluvium ($C_A$), average lowering of the erosion area ($E_M$), and average removal ($A_a$) are listed.

| Phase | Period | $a_a$ (a) | $E_L$ (m) | $C_L$ (m) | $C_M$ (m) | $C_A$ (m$^2$) | $E_M$ (mm) | $A_a$ ($\mathrm{t\,ha^{-1}\,a^{-1}}$) |
|---|---|---|---|---|---|---|---|---|
| 6 | Hallstatt Period | 100 | 63 | 2 | 0.1 | 0.2 | 3 | 0.5 |
| 7 | Roman Empire | 1 | 57 | 7 | 0.12 | 0.8 | 14 | 224 |
| 9.1 | Early Middle Age | 200 | 60 | 5 | 0.1 | 0.5 | 8 | 0.6 |
| 9.2 | Peak Middle Age | 300 | 62 | 17 | 0.2 | 3.4 | 55 | 3 |
| 9.3 | Late Middle Age | 1 | 54 | 25 | 0.12 | 3 | 56 | 900 |
| 9.4 | Late Middle Age | 1 | 39 | 40 | 0.12 | 4.7 | 121 | 1925 |
| 10 | Modern Times | 500 | 45 | 34 | 0.5 | 17 | 378 | 12 |

Today at the convex part of the upper slope the weathered Triassic sandstone lies directly beneath the ploughing horizon. The maximum slope gradient of the convex upper slope was 30% before land use started. Since then, the relief was flattened to the present gradient of 16%. The slope form has also changed. From the earlier convex–concave slope, an almost straight slope with a long stretching middle slope has developed (Figure 7).

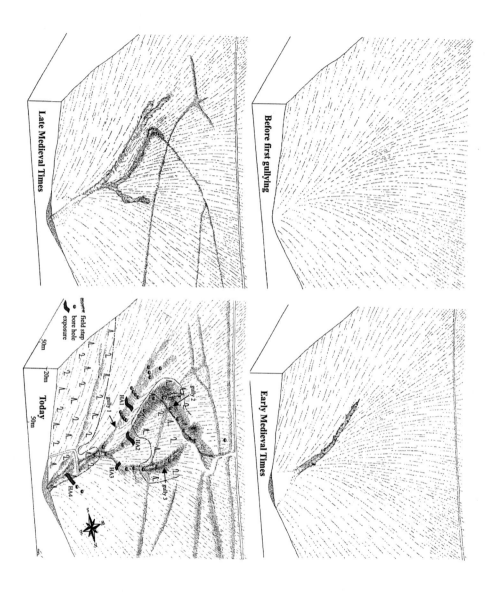

sive land use is assumed at Hainbach for the 13$^{th}$ and the early 14$^{th}$ centuries. In the middle of the 14$^{th}$ century, the village Dörfles was abandoned (Heinold–Fichtner 1951). It is likely, that agricultural land use at the Hainbach valley stopped also at this time. In the following period, dense vegetation covered the surface preventing soil erosion. During the 15$^{th}$ century until the end of the 16$^{th}$ century agricultural land use increased again (Dotterweich 2002). During the Thirty Years War in the first half of the 17$^{th}$ century, the population density of Dörfles and land use intensity in the Hainbach valley decreased dramatically (Dotterweich 2002). The intensity of land use increased again in the 18$^{th}$ century. In the oldest large–scale map from 1853 (Urmesstischblatt NW 102–08, Landesvermessungsamt Bayern), the land use pattern indicates that only small parts of the catchment area were used as fields. According to the land use history, the records of historical rainfall events and the dated sediments, the first incision of the gullies 2 and 3 took place during intensive rainfall events in the first half of the 14$^{th}$ century. The gullies then grew upslope until the 18$^{th}$ century. At the same time, the lower parts of the gully system were filled with sediments. In particular, the clayey, humic and very thin laminated layers are sediments derived from hillslope erosion. At around 1850, the gully system had almost the same appearance as at present. Nowadays the gully system is growing only occasionally at the upper end of gully 2. During heavy rainfalls, water is collected in field furrows and ploughing rills and runs into the gullies.

Thus, the gully system with a total length of 400 m, and a volume of 6430 m$^3$ was established between medieval times and modern times. Gullying dissected 15 % of the surface in the catchment area, i.e. 4400 m$^2$ of 30000 m$^2$.

## 5.3 Landscape dynamic in the region of Kronach

The investigated geoarchives have recorded the landscape development at Catena Friesen and Hainbach. The results obtained cannot easily be extrapolated to the whole region. It is reasonable to assume that the major land use changes, as abandoning villages in the 14$^{th}$ century, and major weather phenomena such as heavy rainfall events, affected the whole Kronach region. The spatial pattern and intensity of soil erosion, however, depend on many other factors as well (size of the catchment area, slope length, slope angle, substrate, and others). Therefore, it is only possible to approximate soil loss values for the Kronach region as a whole. The landscape in the region of Kronach was predominantly wooded from Neolithic times until the end of the Dark Ages. This is supported by the low number of archaeological finds from these periods. Short clearing phases in small areas interrupted the geomorphic stability, and locally soils and sediments were transferred downslope as can be seen at Catena Friesen. Dramatic soil erosion effects due to Middle Age land use changes are evident. For example, up to 30 cm of the soil was truncated at the middle and upper slope at Catena Friesen. The heaviest medieval soil erosion event took place in the first half of the 14$^{th}$ century. Up to 121 mm soil were eroded during a single soil erosion event, probably in July 1342 when the largest rainfall event of the last millennium occurred (Glaser 2000). For the whole region of Kronach, intensive erosion on all agriculturally used slopes is assumed for this time, as was documented for other regions (Bork 1983, 1988, Bork et al. 1998, Dotterweich et al. 2002c). Also gullying occurred: At Hainbach after deforestation surface runoff occurred during heavy rainfall events. First, the natural landform confined the runoff pathways and a gully incised in Early and High Medieval times. In Late Medieval and Modern times the land use structures favoured a concentration of runoff along artificial pathways. Two further gullies were formed during this time. A similar development of other gullied slopes in the region of Kronach is expected. For Modern times, soil erosion at both locations is evident. In contrast to the medieval erosion events, the differentiation of distinct erosion events for Modern times is only possible in the gully fills at Hainbach. There, more than 10 high intensity phases of gullying can be differentiated. It is likely that at Catena Friesen and at many other slopes in arable land soil erosion events occurred at the same times. As shown at Catena Friesen, until the end of the Dark Ages a luvisol had developed. This was probably also the case on many other slopes in the region. In Medieval and Modern times most of the upper soil was eroded and transported downslope. With the soil also nutrients were transported and deposited in low lying, mostly wet areas. These areas are difficult to use for agriculture, while the soils at the upper slopes lost their fertility. At the upper slope of Catena Friesen after only a few erosion events in Medieval times the nutrient poor Triassic bedrock was completely exposed. This was probably the main reason to a surrender of farming activity. Extended village abandoning in the Kronach region in the 14th century was probably caused by the extreme soil erosion events and their results.

## 5.4   The landscape dynamic in northern Bavaria

Detailed information about the frequency and intensity of soil erosion in northern Bavaria are scarce. Only the example Wolfsgraben some kilometres north of Bamberg gives detailed information about the landscape dynamic since Medieval times (Schmitt et al. 2002 and Dotterweich et al. 2002b, 2002c). At this site, intensive hillslope erosion and gullying took place in the 14th century. After this event until the 18th century, more than 20 phases of hillslope erosion removed around 7 cm of soil from the slopes. The sediments were deposited in the medieval gully. In contrast to the gully system at Hainbach, the Wolfsgraben sediments were cut deeply by a second gully in the 18th or 19th century. This is probable due to the different land use patterns at the time. At Hainbach the slopes were mainly used as pasture and forest and thus protecting the surface from further soil erosion. In contrast, the catchment area of the Wolfsgraben was used as arable land, hop–gardens and vineyards and thus allowing surface run–off and soil erosion. In northern Bavaria, more than 80 gully systems similar to the gullies at Hainbach and Wolfsgraben were mapped in an area of 200 km$^2$ on sandy–loamy Triassic substratum (Dotterweich, unpublished). Many of these gullies have partly or completely been filled with sediments containing artefacts from Medieval and Modern times. It is assumed that these gully systems had a similar development as the gully systems at Wolfsgraben or Hainbach. Also most of the slopes investigated by Machann et al. (1970), Abraham de Vazquez et al. (1985), Hildebrandt et al. (1993), Garleff (1988, 1989), Rösner et al. (1999) were formed mainly by few intensive soil erosion events.

The larger the area the greater is also the variability of the factors involved in the soil erosion processes. Land use history patterns and patterns of historic precipitation vary in a larger range. To obtain more detailed information about landscape dynamics in northern Bavaria more case studies of geoarchives in key catchments would be necessary, as well as regional scale studies of land use history and past climate.

# 6   Conclusions

High–resolution analyses of soils and sediments allow a detailed quantitative reconstruction of the intensity and frequency of past soil erosion. Together with information on land use history and extreme weather events, it is possible to explore the causes and dynamics of past soil erosion events and their long–term effects on the interactions between man and environment. During the last 5000 years, land use was the controlling factor of landscape evolution in northern Bavaria. Woodland clearing and subsequent farming allowed soil erosion on fields during intense rainfall events. The highest values of hillslope erosion and the development of many gullies took place in Late Medieval times, during a period with intensive land use and extraordinary heavy rainfalls. It is possible that only few soil erosion events led to a complete removal of the upper soil on many farmed slopes. In the following centuries soil erosion occurred too, but the pathways of surface runoff and the intensity of soil erosion was mainly controlled by land use patterns. Until today, the rainfall events did not reach the intensities of the 14th century. Only in the 18th century, a second period with strongly accelerated soil erosion and gullying took place. In agreement with numerous other results from Central Europe, changes in land use patterns force soil erosion and climatic effects e.g. the higher intensity of rainfall events during this times, are the most probable reasons for the accelerated soil erosion in this time (cf. Bork et al. 1998).

For a better understanding of landscape development, modelling of long–term effects of land use on soil erosion is essential. Missing data about past land use patterns could be derived with statistical methods from the results of archaeological investigations, historical research activities and paleo–botanical analyses. Further case studies of the landscape development of key catchments are also important to obtain more information about the long–term effects of land use on the landscape as well as on socio–economic, demographic, and cultural factors and the interactions between these components.

## Acknowledgements

To all the persons and institutions, which have supported the project, I am very

Mr. Fischer–Weiß and H. Wich for the possibility to dig on their fields and the valuable discussion. G. Förtsch, R. Graf and K. Schäbitz for the valuable discussion. F. Glaßer, S. Reiß, U. Seidel and the Yanmar B25V (the excavator) for their support in the fieldwork. P.M. Grootes and H. Erlenkeuser for the Radiocarbon Dating and B. Mauz for the OSL–Dating. G. Klose for the hand drawing of some pictures. G. Schmidtchen and H.–R. Bork for their help in the field, the valuable discussions and the comments on the text. I express my thanks to J. Haberstroh and H. Endres for their support in forming archaeological objectives and dating of the potsherds. I thank the Cusanuswerk for the financial support making the entire research project possible.

# References

Abraham de Vazquez, E.M., Garleff, K., Schäbitz, F., and Seemann, G. (1985): Untersuchungen zur vorzeitlichen Bodenerosion im Einzugsgebiet des Ellerbaches östlich Bamberg. LX. Ber. Naturforsch. Ges. Bamberg: 173–190.

AG Boden (1994): Bodenkundliche Kartieranleitung. Arbeitsgruppe Boden Hrsg. von der Bundesanstalt für Geowissenschaften und Rohstoffe und den Geologischen Landesämtern in der Bundesrepublik Deutschland. 4. verb. u. erw. Aufl., Hannover.

Aitken, M.J. (1985): Thermoluminescence Dating. Academic Press, London.

Becker, B. (1983): Postglaziale Auwaldentwicklung im mittleren und oberen Maintal anhand dendrochronologischer Untersuchungen subfossiler Baumstammablagerungen. Hannover, Geolog. Jahrbuch, A 71: 45–59.

Birks, H.J.B., Line, J.M., and Persson, T. (1988): Quantitative Estimation of Human Impact on Cultural Landscape Development.–In: Birks, H., Birks, H.J.B., Kaland, P.E., and Moe, D. (eds.), The Cultural Landscape – Past, Present, Future. Cambridge University Press, Cambridge, pp. 229–239.

Bork, H.–R. (1983): Die holozäne Relief– und Bodenentwicklung in Lössgebieten.–In: Bork, H.–R., and Ricken,W. (eds.): Bodenerosion, Holozäne und Pleistozäne Bodenerntwicklung, Catena Supplement 3, Braunschweig.

Bork, H.–R., (1988): Bodenerosion und Umwelt. Landschaftsgenese und Landschaftsökologie 13, Braunschweig.

Bork, H.–R., Bork, H., Dalchow. C., Faust, B., Piorr, H.–P., and Schatz, T. (1998): Landschaftsentwicklung in Mitteleuropa. Klett–Perthes, Gotha.

Dannapfel, M. (1991): Kurzer Abriss der Geologie von Franken. Bayreuther Bodenkundliche Berichte 17: 11–19.

Demattio, H., (1998): Kronach, der Altlandkreis. Kommission für Bayerische Landesgeschichte, München.

Dotterweich, M. (2002): Landnutzungsbedingte Kerbenentwicklung während Mittelalter und Neuzeit in der Obermainischen Bruchschollenlandschaft bei Kronach.–In: Bork, H.–R., Schmidtchen, G., and Dotterweich, M. (eds.): Bodenbildung, Bodenerosion und Reliefentwicklung im Mittel– und Jungholozän. Forschungen zur deutschen Landeskunde (in press).

Dotterweich, M., Haberstroh, J., and Bork, H.–R. (2002a): Mittel– und jungholozäne Siedlungsentwicklung, Landnutzung, Bodenbildung und Bodenerosion an einer mittelalterlichen Wüstung bei Friesen, Landkreis Kronach in Oberfranken.–In: Bork, H.–R., Schmidtchen, G., and Dotterweich, M. (eds.): Bodenbildung, Bodenerosion und Reliefentwicklung im Mittel– und Jungholozän. Forschungen zur deutschen Landeskunde (in press).

Dotterweich, M., Schmitt, A., and Bork, H.-R. (2002b): Jungholozäne Bodenerosion und Kerbenentwicklung im Wolfsgraben bei Bamberg.-In: Bork, H.-R., Schmidtchen, G., and Dotterweich, M. (eds.): Bodenbildung, Bodenerosion und Reliefentwicklung im Mittel- und Jungholozän. Forschungen zur deutschen Landeskunde (in press).

Dotterweich, M., Schmitt, A., Schmidtchen, G., and Bork, H.-R. (2002c): Quantifying historical gully erosion in northern Bavaria. Catena (in press).

Emmert, U., and v. Horstig, G. (1972): Geologische Karte von Bayern 1:25000, Erläuterungen zum Blatt Nr. 5734 Wallenfels. Bayerisches Geologisches Landesamt, München.

Endres, R. (1990): Die Folgen des 30-jährigen Krieges in Franken. Mitteilungen der Fränkischen Geographischen Gesellschaft 35/36, Erlangen.

Garleff, K. (1987): Radiokarbondaten aus den Talauensedimenten des Ellerbaches östlich von Bamberg. LXII. Bericht Naturforschende Gesellschaft Bamberg: 179–184

Geologische Karte von Bayern, 1972. Blatt 5734 Wallenfels, München.

Gerlach, R. (1990): Flussdynamik des Mains unter dem Einfluss des Menschen seit dem Spätmittelalter. Trier, Forschungen zur deutschen Landeskunde, Band 234.

Glaser, R. (1991): Klimarekonstruktion für Mainfranken, Bauland und Odenwald anhand direkter und indirekter Witterungsdaten seit 1500. Akademie der Wissenschaften und der Literatur, Paläoklimaforschung 5, Mainz, Stuttgart, New York.

Glaser, R. (2001): Klimageschichte Mitteleuropas – 1000 Jahre Wetter, Klima, Katastrophen. Wissenschaftliche Buchgesellschaft Darmstadt.

Heinold–Fichtner (1951): Die Bamberger Oberämter Kronach und Teuschnitz. Schriften des Instituts für fränkische Landesforschung an der Universität Erlangen, Band 3, Erlangen.

Hildebrandt, H., and Kauder, B. (1993): Wüstungsvorgänge im westlichen Steigerwald. Forschungskreis Ebrach e.V., Ebrach.

Küster, H.-J. (1998): Geschichte des Waldes. C.H. Beck München.

Machann, R., and Semmel, A. (1970): Historische Bodenerosion auf Wüstungsfluren Deutscher Mittelgebirge. Geographische Zeitschrift, 58. Jahrgang: 250–266.

Mauz, B., Lang, A., and Dikau, R. (2002): Optical dating of colluvia and guly-fill sediments using quartz.-In: Bork, H.-R., Schmidtchen, G., and Dotterweich, M. (eds.): Bodenbildung, Bodenerosion und Reliefentwicklung im Mittel- udn Jungholozän. Forschungen zur deutschen Landeskunde.

Munsell (1975): Soil Color Charts. US Department of Agriculture, Baltimore, Maryland.

Rösner, U., and Töpfer, C. (1999): Historische Bodenerosion auf Flurwüstungen im westlichen Steigerwald. Mitteilungen der Fränkischen Geographischen Gesellschaft Erlangen, Band 46: 27–73.

Sage, W. (1996, ed.): Oberfranken in vor- und frühgeschichtlicher Zeit. Bayerische Verlagsanstalt Bamberg, 2. Ed.

Schirmer, W. (1983): Die Talentwicklung an Main und Regnitz seit dem Hochwürm.-In: Schirmer, W. (ed.): Holozäne Talentwicklung – Methoden und Ergebnisse. Geol. Jb. A 71: 11–43.

Schirmer, W. (1988): Junge Flussgeschichte des Mains um Bamberg. Hannover, DEUQUA, 24. Tagung, Exkursion H: 1–39.

Schmitt, A., Dotterweich, M., Schmidtchen, G., and Bork, H.-R. (2002): Vineyards, hopgardens and recent afforestation: effects of late Holocene land use change on soil erosion in northern Bavaria, Germany. Catena (in press).

Schönwiese, Ch. (1995): Klimaänderungen. Daten, Analysen, Prognosen. Springer Verlag.

Stöhr, Coelestinus und Hieronymus (1825): Neue Chronick der Stadt Cronach. Cronach.

Stuiver, M., Reimer, P.J., Bard, E., Beck, J.W., Burr, G.S., Hughen, K.A., Kromer, B., McCormac, F.G., Plicht, J., and Spurk, M. (1998): INTCAL98 Radiocarbon age calibration 24,000 – 0 cal BP. Radiocarbon, 40:1041–1083.

v. Geldern–Crispendorf (1930): Kulturgeographie des Frankenwaldes.–In: Beih. z. d. Mitt. d. Sächs.–Thür. Ver. f. Erdk. z. Halle a. S. Nr. 1.

Weikinn, C. (1958–1963): Quellentexte zur Witterungsgeschichte Europas von der Zeitenwende bis zum Jahre 1850. Bd. I–IV. Berlin

Züchner, Ch. (1996): Die Steinzeit in Oberfranken.–In: Sage, W. (ed.): Oberfranken in vor– und frühgeschichtlicher Zeit. Bayerische Verlagsanstalt Bamberg, 2. Aufl.: 25–64.

# Quantification of past soil erosion and land use / land cover changes in Germany

Hans–Rudolf Bork[1] and Andreas Lang[2]

[1] Ökologie–Zentrum der Christian–Albrechts–Universität zu Kiel, Olshausenstrasse 40, 24098 Kiel, Germamy
[2] Fysische en Regionale Geografie, K.U. Leuven, Redingenstraat 16, 3000 Leuven Belgium

**Abstract.** Knowledge on land use changes dating back to the period of Neolithic to High Middle Ages in Germany is still rather limited. Although numerous local archaeological and palynological records exist, regional scale reconstructions are rare. Information on past soil erosion available from case studies on soil translocation and soil formation has also only rarely been regionalised. Here, we review approaches aiming at integrating results from case studies at different locations, in order to establish a regional history of soil erosion in prehistoric and historic Germany.

It can be shown that soil erosion is not a modern problem. During Neolithic through to medieval times, soil erosion occurred more frequently during phases of stronger human impact. The oldest soil erosion derived sediments were deposited in early Neolithic times. Widespread soil erosion is found for the Bronze Age, Iron Age and Roman periods, with maximum rates occurring in the medieval period. This indicates that soil erosion during these periods was triggered mainly by the intensity of land use, suggesting that this was the critical factor for the landscape's sensitivity to erosion. For the later medieval and modern periods - when the area used as arable land reached more or less its present extent - maxima in soil erosion can be associated with high magnitude rainfall events. Extreme soil loss occurred during the first half of the 14th century and - less pronounced - in the second half of the 18th century.

## 1 Introduction

One of the most important questions in palaeoecological research focusing on the whole period of agriculture in Central Europe is: When and to which extent have woodlands been transformed to arable land and pasture? Despite the fact that archaeological findings can give important hints to when and where transformation took place, many uncertainties still remain. For example, although the vegetation composition, i.e. the occurrence of certain species in an area and during a specific period of time can clearly be derived from palynological findings, the extent of the region represented by such findings is difficult to define. Furthermore, they only show the existence of a species but do not allow quantifying vegetation patterns (Beug 1992, Küster 1996). Other sources that have been used to obtain information on land use change and settlement history are the ages of villages and towns. This allows establishing periods of settlement foundation but the degree of land cover change and soil transformation during such a period cannot be inferred. Even written documents from Early and High Medieval Times usually do not contain information on the spatial extent of a specific land use type and the contemporary state of the soils.

Hence, despite the wealth of archaeological, historical and palynological data our knowledge about land use changes for the period from the Neolithic to High Medieval times is still rather limited. There are numerous local archaeological and palynological records, but only in few cases extrapolations to larger areas were successful (Firbas 1949, 1952, Lange 1971, 1989, Gringmuth-Dallmer 1983, 1989, Beug 1992, Küster 1996).

Since Late Medieval times, written documents represent an important source of information for local and, in some cases, also for regional land use changes (e.g. Jäger 1958, 1963, Abel 1976, 1978, Born 1980, Schenk 1996). But only since the 19th century data on changes of woodlands, arable land, and pasture as well as soil dynamics are more ubiquitous.

A key to reducing such deficits is extracting information from soil formation and soil translocation processes, either for specific sites (spots), for catenas and small catchments, or for whole regions. Soils preserve information about the conditions of their formation, whereas information on the causes and the degree of soil truncation is recorded in soil erosion derived sediments. Extracting this information is, however, not an easy task. Often, soils are polygenetic, i.e. the cumulative result of several different climatic and land use periods. It is, therefore, important to investigate soil sequences along slopes. At some parts of the slope  especially in accumulation areas at concave lower slopes  soils resulting from different soil forming periods may be recognisable. At other locations, for example at middle and upper slopes, soil–forming periods cannot be differentiated because the same substratum underwent different processes of pedogenesis in all formation periods and younger ones overprinted earlier transformations.

Soil development is, among other controlling factors such as landform, parent rock, and local climate, dependent on vegetation cover and land use. Hence, soils that developed under deciduous forest and pine forest are completely different from soils that formed under cultivated land. Additionally in Central Europe soil erosion occurs on arable land: Due to insufficient vegetation cover the soil is susceptible to erosion. This can be either initiated by rainfall events (water erosion) and storms (wind erosion), or by ploughing and other farming practices (tillage translocation). Before being deposited eroded soil particles are transported over a certain distance depending on landform, vegetation, and magnitude of a runoff event: Particles transported by cultivation will be deposited within a field parcel. Particles transported by water during low and medium magnitude rainfall events are mainly deposited at concave lower slopes, in shallow zero order basins, and in small alluvial fans. During high magnitude rainfall events different processes are operating and eroded particles are often evacuated from the slopes and transported to the rivers (Lang et al., in print).

# 2   Reconstructing past changes in soils and land use

As a first step in a reconstruction approach, soil profiles and their evolution are investigated in terms of their spatial and temporal context. Study sites should be chosen along the flow paths of surface and subsurface flows. In a second step, the site results can then be integrated into catenas stretching from the drainage divide to the thalweg.

Simple examples clearly illustrate how small differences in soil evolution can be used for reconstructing phases of vegetation cover and land use: After only a few years under pasture Ranker and Rendzina soils develop with a humic horizon of 1 to 3 cm depth. Under the same conditions the development of Ranker and Rendzina with humic horizons of 10 to 20 cm depth, however, takes several decades or even few centuries. Several centuries of woodland coverage are necessary for the formation of cambisols and luvisols. Podsols, on the other hand, form after a few centuries in a substratum low in nutrients and under a heath cover. One single ploughing event alone can cause the development of a clear ploughing horizon.

Nevertheless, conclusions on former vegetation cover and land use from soil type and related characteristics can only be drawn, if (I) sediment characteristics can clearly be recognised from soil characteristics and if (II) an accurate chronology can be established for the different processes involved. (I) can be rather difficult, for example, if an autochthonous Bv-horizon which formed under woodland has to be distinguished from sediment derived from a Bv-horizon that was reworked on arable land. Still, in many cases differentiation is possible even in the field. A careful description and analysis of a structures 3D shape should reveal the difference: An autochthonous Bv-horizon shows a gradual transition to the underlying C-horizon, whereas a sediment formed from a reworked Bv-horizon shows a distinct and often sharp lower boundary. Additionally, reworked Bv-horizons often show sedimentary structures such as fine lamina that can often be identified in the field. In the lab, micromorphological techniques can help identifying such features. Finally, phases of sedimentation and soil formation can be stratigraphically ordered and their evolution can be described in chronological sequence. (II) Numeric chronologies can usually only be derived for sediments not for soil forming processes. Few techniques allow direct dating of sediment deposition in such environments and most studies relay on indirect age information. Indirect information is obtained mainly from ages of objects incorporated in the sediment (archaeological classification of artefacts or [14]C–dating of charcoal). Direct dating is possible with optically stimulated luminescence (OSL) -dating, or when organic material produced in situ is radiocarbon dated. Examples of case studies using such approaches are presented by Dotterweich (*this volume*) and Schmidtchen and Bork (*this volume*).

# 3   Regionalisation of soil erosion and land use changes

## 3.1   Middle Ages and modern times in Germany

Using approaches similar to the above H. -R. Bork and co-workers investigated
more than 2200 locations in Germany since 1978 (e.g. Bork 1983, 1988, Bork et
al. 1998). Based on these results an initial attempt towards a regionalisation was
carried out. Originally, the locations were chosen according to the specific needs
of different research projects. Thus, the spatial distribution of the locations is
neither homogeneous nor random and a simple extrapolation of the statistical
mean of the results is not valid. Thus, as a first step towards regionalisation
clusters of locations were formed based on homogeneous regions with largely
similar substratum, soil evolution and soil translocation history. In the south-
eastern part of Lower Saxony, Germany, for example, representative values of
soil erosion rates were calculated on the basis of results from ca. 800 locations
in an area of $280\,\mathrm{km^2}$ (Bork 1983). First, in a hierarchical approach, a few $\mathrm{km^2}$
large landscape elements were chosen randomly. Second, within each element
catenas were selected, again randomly. Subsequently, for each catena selected
the following values were calculated:

1. the volume of eroded soil from the amount of profile truncation due to medieval and modern soil erosion
2. the volume of sediment at the lower slope and in the floodplain due to this
   erosion
3. sediment loss to the channel as the difference of the first two values.

Each set of data was considered representative for the specific landscape
element. Finally, the mean values were calculated for all landscape elements to
represent the whole region of $280\,\mathrm{km^2}$. Results show that at the upper and middle
slopes an average of $2.3\,\mathrm{m}$ soil were eroded since Early Medieval Times. More
than 80% of this material was not transported out the catchment but deposited
on the lower slopes where it still rests today and can now be used as geoarchive.

But reconstruction of soil and land use changes did not only provide values
for total erosion rates from Medieval to Modern times in southeast Lower Saxony.
In a more detailed analysis, data for single land use periods could be determined,
allowing the identification and quantification of single extreme events, e.g. in the
first half of the 14th century (Bork 1988, Bork et al. 1998). However, these results
cannot be regarded as being representative of other areas, since similar values
for soil erosion have only been reported from few regions in Central Europe, as
for example the Kraichgau hills in South West Germany (Lang et al. 1999).

A different approach was used in eastern Brandenburg. Rather than using
a random distribution, catenas to be studied were selected based on field experience according to how typical they are for the study area. Additionally, two
steep catchments in which a gully system has developed were investigated. Again
detailed soil and sediment stratigraphies were established and allowed differentiation of more than 50 evolutionary steps in the period from Medieval to Modern
times. Dating of the sediments allowed calculation of erosion rates for specific

land use periods that were reconstructed from historic documents. Rather surprisingly, results of specific extreme events could be identified in the sediments. As in south-eastern Lower Saxony, the events dated back to the 14th century. Due to the different approach in the east Brandenburg study, however, it was more difficult to define representative study sites, and hence, to determine spatial mean values for soil erosion and land use change. To be considered representative, a study area had to meet specific topographical, geological and pedological criteria. It showed that some of the study sites (catenas) were characterised by above–average inclination and below-average slope length, which had to be taken into account for the calculation of the spatial averages of soil erosion rates. For eastern Brandenburg, from Medieval to Modern times a mean soil loss of 0.5 m could be determined, the main part of which occurred in the first half of the 14th century.

This approach was applied to other German landscapes and their spatial averages for soil erosion and land use change - based on samples of different sizes and obtained by different means - were subsequently used equally. Despite the large number of 2200 study sites, for some German landscapes values for soil erosion had to be estimated. Estimates were derived using a knowledge-based approach rather than simple interpolations and were based on similarities in certain characteristics such as landform, and substratum, as well as likely land use history. Areas with clearly different character, the Alps, were excluded from the analysis.

Results of this regionalisation attempt are given in Figure 1. Changes in land cover/land use and estimated rates of soil erosion are plotted versus time for the period since the Early Middle Ages.

## 3.2  South Germany before the Middle Ages

For periods earlier than the Middle Ages  as usual when going further back in time  the quantity and quality of information is even further reduced. Written records are largely missing for Germany and even if they are available they should be treated with caution. But also soil and sediment based approaches as above have to face more challenges: the record is less and less complete as earlier periods are envisaged. The chance of older sediments being eroded is higher, as is the risk of a total overprint of traces of earlier soil formations. Also, chronometric information is harder to obtain. Lang and Hoenscheidt (1999) clearly show how due to reworking of artefacts and organic remains indirect dating approaches can be misleading.

During the past few years, numerous local studies have been carried out in the loess hills of South Germany. Detailed information on soil erosion and sediment storage exists especially from the surroundings of archaeological excavation sites. However, from most of these studies only stratigraphic and chronological information is available. Volumes of erosion and deposition have not been determined. Thus, the extrapolation of findings from local case studies to a more regional scale is problematic. A first graphical approach for a frequency analysis of phases of soil erosion is presented by Lang (2002). OSL (optically stimulated

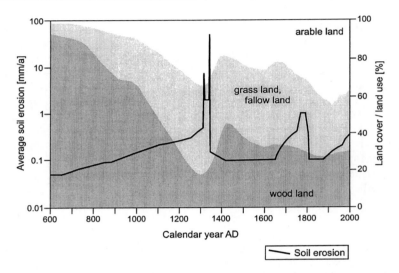

**Fig. 1.** Land cover/land use and soil erosion in Germany (except for the Alps) since the Early Middle Ages (data from Bork et al., 1998). The average soil erosion in mm/a is plotted as solid line (left scale). For three land use classes the proportion of land cover is plotted as grey tints (right scale).

luminescence) ages of soil erosion derived colluvial sediments were analysed for the period from the beginning of agriculture (the Neolithic) until 1200 AD. The frequency distribution was constructed by: (1) representing the optical ages by Gaussian-distributions and (2) summing up all the single curves (figure 2). A first significant increase in colluviation occurred during the Bronze Age. During the Iron Age/Roman period and at around 800 AD distinct maxima appear in the distribution and highest frequencies are present towards the end of the period analysed at around 1100 AD. It is therefore concluded that phases of increased colluviation occur during periods of stronger human impact.

Conclusions that can be drawn from such an approach are restricted by the still rather limited amount of data, sampling bias and other factors. However, for the period analysed, and on a regional and long–term scale, colluviation seems to be mainly triggered by the intensity of land use. Climatic fluctuations are of secondary importance only. Rainfall events sufficiently erosive to produce colluvial sediments occurred throughout all agricultural periods. The limiting factor for colluvium production was the sensitivity of the landscape to erosion – which ultimately is controlled by land use.

## 4  Conclusions

The results presented above reflect the present state of the art. The integration of results from ongoing and future work will allow further expansion and refinement

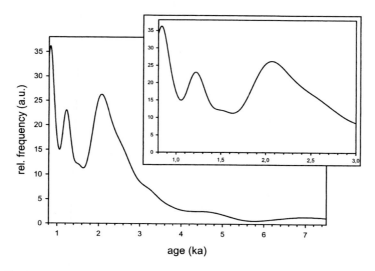

**Fig. 2.** Probability density distribution of 60 OSL-ages for soil erosion-derived colluvium from South Germany for the period 0.8 ka to 7.5 ka. Inset: Enlargement for the period 0.8 ka to 3.5 ka (after Lang, 2002).

of the history of soil erosion in Germany and Central Europe. Nevertheless, it has become clear that soil erosion is not a modern problem in Germany. For the younger period (medieval and modern times) and due to the higher resolution of data, the soil erosion can be related to high magnitude rainfall events. Extreme maxima of soil erosion occurred during the first half of the 14th century a period during which extreme rainfalls coincided with the all time low in woodland cover in Germany. A second, but less pronounced period is evident during the second half of the 18th century. This again is a period for which documentary evidence of extreme rainfalls and high runoff exists.

Regarding soil erosion from Neolithic to Medieval times in the loess hills of South Germany - due to the lower precision of the data - it can only be stated that colluviation occurred more frequently during phases of stronger human impact. The oldest colluvial sediments were deposited during early Neolithic times. The distribution shows a second small maximum towards the end of the Neolithic period. Many colluvia develop in the Iron Age and Roman periods, while the maximum number of optical ages relate to medieval times. This indicates that colluviation during this period was dominated by the intensity of land use. Climatic fluctuations seem to play a secondary role, considering that sufficiently erosive rainfall events occurred during all agricultural periods. Probably the critical factor was the landscapes sensitivity to erosion.

# References

Abel, W. (1976): Die Wüstungen des ausgehenden Mittelalters. Quellen u. Forschungen z. Agrargeschichte 1: 186 p., Stuttgart (Fischer).

Abel, W. (1978): Geschichte der deutschen Landwirtschaft vom frühen Mittelalter bis zum 19. Jahrhundert. Deutsche Agrargeschichte II: 370 p., Stuttgart (Ulmer).

Beug, H.–J. (1992): Vegetationsgeschichtliche Untersuchungen über die Besiedlung im Unteren Eichsfeld, Landkreis Göttingen, vom frühen Neolithikum bis zum Mittelalter. Neue Ausgrabungen und Forschungen in Niedersachsen 20: 261-339, Hildesheim.

Bork, H.–R. (1983): Die holozäne Relief– und Bodenentwicklung in Lößgebieten – Beispiele aus dem südöstlichen Niedersachsen.– In: H.–R. Bork, and W. Ricken (eds.): Bodenerosion, holozäne und pleistozäne Bodenentwicklung. Catena Suppl. 3: 1-93.

Bork, H.–R. (1988): Bodenerosion und Umwelt. Landschaftsgenese und Landschaftsökologie 13: 249 p., Braunschweig (Techn. Univ. Braunschweig).

Bork, H.–R., Bork, H., Dalchow, C., Faust, B., Piorr, H.–P., and Schatz, T. (1998): Landschaftsentwicklung in Mitteleuropa. 328 p., Gotha (Perthes).

Born, M. (1980, ed.): Siedlungsgenese und Kulturlandschaftsentwicklung in Mitteleuropa. Gesammelte Beiträge, herausgegeben von K. Fehn.- Geogr. Zeitschr. Beihefte. Erdkundl. Wissen 53: 528 p., Wiesbaden (Steiner).

Dotterweich, M. (2003): Land Use and Soil Erosion in northern Bavaria during the last 5000 Years. In: Lang, A., Hennrich, K.P., and Dikau, R. (eds.): Long term hillslope and fluvial system modelling: Concepts and case studies from the Rhine river catchment, Lecture Notes in Earth Sciences, Springer, Heidelberg: 203–232.

Firbas, F. (1949): Spät- und nacheiszeitliche Waldgeschichte Mitteleuropas nördlich der Alpen. 1. Band: Allgemeine Waldgeschichte, 480 p., Jena (Fischer).

Firbas, F. (1952): Spät- und nacheiszeitliche Waldgeschichte Mitteleuropas nördlich der Alpen. 2. Band: Waldgeschichte der einzelnen Landschaften, 256 p., Jena (Fischer).

Gringmuth-Dallmer, E. (1983): Die Entwicklung der frühgeschichtlichen Kulturlandschaft auf dem Territorium der DDR unter besonderer Berücksichtigung der Siedlungsgebiete. Schriften z. Ur- u. Frühgeschichte 35: 166 p., Berlin (Akademie-Verl.).

Gringmuth-Dallmer, E. (1989): Landwirtschaft und Landesausbau in den germanisch-deutschen Gebieten vom 8.-13. Jh.– In: Herrmann, J. (ed.): Archäologie in der Deutschen Demokratischen Republik, Denkmale und Funde 1, Archäologische Kulturen, geschichtliche Perioden und Volksstämme: 238-248.

Jäger, H. (1958): Entwicklungsperioden agrarer Siedlungsgebiete im mittleren Westdeutschland seit dem frühen 13. Jahrhundert. Würzburger Geogr. Arb. 6: 136 p., Würzburg.

Jäger, H. (1963): Zur Geschichte der deutschen Kulturlandschaften. Geogr. Zeitschr. 51: 90-143.

Küster, H. (1996): Geschichte der Landschaft in Mitteleuropa. Von der Eiszeit bis zur Gegenwart. 424 p., München (C.H.Beck'sche Verlagsbuchhdlg.).

Lang, A. (2002): Phases of soil erosion–caused colluviation in the loess hills of South Germany. Catena (in print).

Lang, A., and Hönscheidt, S. (1999): Age and source of soil erosion derived colluvial sediments at Vaihingen-Enz, Germany. Catena, 38(2): 89–107.

Lang, A., Kadereit, A., Behrends, R.H., and Wagner, G.A. (1999): Optical dating of anthropogenic sediments at the archaeological excavation site Herrenbrunnenbuckel, Bretten-Bauerbach, Germany. Archaeometry, 41(2): 397–411.

Lang, A., Bork, H.-R., Mäckel, R., Preston, N., Wunderlich, J., and Dikau, R.: Changes in sediment flux and storage within a fluvial system – some examples from the Rhine catchment. Hydrological Processes (in print).

Lange, E. (1971): Botanische Beiträge zur mitteleuropäischen Siedlungsgeschichte. Schriften z. Ur- und Frühgeschichte 27: 183 p.; Berlin (Akademie-Verlag).

Lange, E. (1989): Aussagen botanischer Quellen zur mittelalterlichen Landnutzung im Gebiet der DDR.– In: Herrmann, B. (ed.): Umwelt in der Geschichte: 26–39.

Schenk, W. (1996): Waldnutzung, Waldzustand und regionale Entwicklung in vorindustrieller Zeit im mittleren Deutschland. Erdkundl. Wissen 117: 326 p., Stuttgart (Steiner).

Schmidtchen, G., and Bork, H.-R. (this volume): Changing Human Impact during the Period of Agriculture in Central Europe: The Case Study Biesdorfer Kehlen, Brandenburg, In: Lang, A., Hennrich, K.P., and Dikau, R. (eds.): Long term hillslope and fluvial system modelling: Concepts and case studies from the Rhine river catchment, Lecture Notes in Earth Sciences, Springer, Heidelberg: 185–202.

# Index

Note: Page references in *italics* refer to figures; those in **bold** refer to tables.

# Lecture Notes in Earth Sciences

For information about Vols. 1–24
please contact your bookseller or Springer-Verlag